W9-BDM-847

TOO HOT TO HANDLE?

TOO HOT TO HANDLE?

SOCIAL AND POLICY ISSUES IN THE MANAGEMENT OF RADIOACTIVE WASTES

Edited by
CHARLES A. WALKER,
LEROY C. GOULD, and
EDWARD J. WOODHOUSE

YALE UNIVERSITY PRESS
New Haven and London

Designed by Nancy Ovedovitz and set in Baskerville type by P & M Typesetting, Incorporated.
Printed in the United States of America by The Alpine Press, Inc., Stoughton, Mass.

Library of Congress Cataloging in Publication Data

Main entry under title:
Too hot to handle?
Includes bibliographies and index.
Contents: The radioactive waste management problem / Leroy C. Gould—Science and technology of the sources and management of radioactive wastes / Charles A. Walker—Nuclear waste management and risks to human health / Jan A. J. Stolwijk—[etc.]
1. Radioactive waste disposal—Addresses, essays, lectures. I. Walker, Charles A. (Charles Allen), 1914– . II. Gould, Leroy C. III. Woodhouse, Edward J.
TD898.T66 1983 363.7′28 82-20000
ISBN 0–300–02899–7
ISBN 0–300–02993–4 pbk.

10 9 8 7 6 5 4 3 2 1

To
Allen,
Anne,
Anthony,
Daniel,
David,
Ellen,
Ilya,
Jill,
John,
Juli,
Lauren,
Laurence,
Maya,
Michelle,
Rebecca,
Richard,
Scott,
Scott,
Steven,
and
Timothy

CONTENTS

FOREWORD

On July 16, 1945, on a desert in New Mexico, the United States exploded the world's first atomic bomb and thereby catapulted the world into the nuclear age. Robert Oppenheimer, director of the Los Alamos project that developed the bomb, later recalled what went through his mind on that fateful day—two lines from the *Bhagavad-Gita* in which God is speaking: "I am become death, the shatterer of worlds; Waiting that hour that ripens to their doom."

Ten years later, in part as appeasement to the bomb's potential for destruction, the United States launched the "atoms for peace" program designed to harness the awesome power of the atom for constructive purposes. "The atom stands ready to become man's obedient, tireless servant," President Eisenhower observed in announcing the new program, "if man will only allow it" (*New York Times,* August 9, 1955, p. 8).

Another generation has now passed and five nations, besides the United States, have developed and tested atomic weapons and nineteen nations have installed nuclear-electric power generating facilities. It has yet to be decided, however, whether the atom will ripen to man's doom or become his obedient, tireless servant. That fate, moreover, depends not only on whether the nations of the world can avoid nuclear war, but also on whether the world's nuclear power plants prove to be safe and whether safe procedures can be developed to handle, and eventually dispose of, the radioactive waste products that issue daily from the world's nuclear arms and nuclear power industries.

As Alvin Weinberg (*Science* 177:27–34), one of the founders of nuclear technology and past director of the Atomic Energy Commission, summed it up:

> we seem to have struck a Faustian bargain. We are given the miraculous nuclear fire . . . as a means of producing very clean and . . . inexhaustible energy. The price that we must pay for this great boon is a

vigilance that in many ways transcends what we have ever had to maintain: vigilance and care in operating these devices, and creation, and continuation into eternity, of a cadre or priesthood who understand the nuclear systems, and who are prepared to guard the wastes. To those of us whose business it is to supply power here and now, such speculations about 100,000 year priesthoods must strike an eerie and unreal sound. . . . But the immediate concern for vigilant, intelligent, and responsible operation of nuclear power plants is not theoretical or remote: It is a heavy responsibility that everyone in the utility industry, public and private, must assume.

Large quantities of highly radioactive wastes have accumulated in the past thirty years as a by-product of commercial production of nuclear-electric power and military production of nuclear weapons. Wastes from commercial power plants include more than eight thousand tons of highly radioactive spent fuel assemblies stored in water-cooled basins at reactor sites and many tons of less radioactive uranium mill tailings, contaminated equipment, discarded clothing, and other materials that have been buried in shallow-earth repositories. Wastes from weapons production include approximately eighty million gallons of highly radioactive solutions and sludges stored in steel tanks and a large quantity of less radioactive wastes in shallow-earth disposal. The military-related wastes are greater in volume than the commercial wastes, although the total radioactivity of materials from the two sources is approximately equal. The inventory of commercial wastes is increasing more rapidly than that of military-related wastes.

Responsibility for vigilance and care in nuclear matters has rested, in the past, almost exclusively with the military, the utility industry, and those government bodies that license and regulate nuclear power producers and military programs. Increasingly, however, owing in part to a general and growing concern about environmental and consumer issues but also to highly publicized accidents at commercial and military nuclear facilities, members of the general public have come to question the management of the nuclear industry and to demand a more direct voice in deciding issues of nuclear safety.

Given the complex and highly technical nature of nuclear power, it is hard for members of the general public to participate wisely in decisions about nuclear safety. For this reason, some have argued that the public should not be involved directly, but instead should rely on technical experts and elected representatives to make the best decisions for them. Others, however, noting how much the public stands to lose if a major nuclear accident should occur, contend that it is not only proper, but essential, that citizens in a democracy have a direct voice in matters as important to their lives as nuclear power.

The purpose of this book is to provide information about one aspect of nuclear safety—the management of radioactive wastes. In particular, the book is designed to assist the reader in understanding the following topics: the history of radioactive waste management in this country and the role of nuclear energy in the future of the United States; the science and technology of the processes that produce radioactive wastes and of methods proposed for managing them; the biological effects of radiation; public attitudes about nuclear power; the nature of risks resulting from technological developments and ways of managing them; and the political institutions and processes that govern radioactive waste management. We have restricted our discussion to problems of radioactive waste management rather than discussing nuclear power in general, not because the latter is not equally, if not more, important, but because the former has received less attention than it deserves and because important decisions about radioactive waste management, decisions that could affect profoundly the future course of nuclear power development and public safety, are right now in the process of being made. We make no pretense that this volume contains all the information that people could use to respond wisely to the issues involved in the management of radioactive wastes. Our hope, simply, is that the volume contains enough information for interested readers to understand the major issues in the debate. The references included in the text refer both to technical documents and to publications that should also be useful to nontechnical readers.

As a group, the authors of this volume are neither for nor against nuclear power. This does not mean, of course, that we do not have individual opinions about whether or not the United States should continue its nuclear program; it means only that we have agreed that the purpose of this work is to inform public opinion rather than to take a personal stand on the nuclear question. We all agree that the safe management of radioactive waste products is an issue that exists right now. The wastes are here today in temporary storage awaiting decisions to determine whether or not they will be reprocessed and whether they will be sequestered in retrievable or irretrievable form in more permanent repositories. Decisions made in the next few years about how to manage radioactive wastes will affect the future course of nuclear power development just as commitments to increased or decreased nuclear power development will inevitably affect the magnitude of the radioactive waste management problem. Nevertheless, the two issues can be separated, in part, and to that extent we have tried to keep our own biases on nuclear power out of our discussion of radioactive waste management. Ultimately, we think, the nuclear question is an issue for the whole country, perhaps the whole world, to decide. We hope that the decisions are made wisely, after careful consideration of the risks and benefits involved.

This book has its origins in discussions held during a workshop-seminar on energy and the social sciences in Yale's Institution for Social and Policy Studies. In that workshop-seminar members of the Yale faculty and research staff meet to discuss social and political implications of energy policy. Representatives of the United States Department of Energy participated for three years, and representatives from industry (General Electric, Northeast Utilities, and Conoco) have attended on a regular and continuing basis. The subject of radioactive waste management is only one of numerous topics discussed by the group, but it proved interesting enough to lead four members of the group (Leroy C. Gould, Jan A. J. Stolwijk, Charles A. Walker, and Edward J. Woodhouse) to undertake the writing and editing of this book. They were fortunate in being able to persuade colleagues in other institutions to contribute discussions of public attitudes (Chapter 4) and perceived and acceptable risks (Chapter 5).

We, the editors, acknowledge with thanks the significant contribution of our fellow authors. Many others have encouraged us and contributed to the discussions that inspired this volume. Among them are William R. Burch, Donald DeLuca, Leonard Doob, Charles E. Lindblom, Guy Orcutt, and Richard R. Nelson of Yale; Fred Abel, Michael J. Biallas, Paul H. Gerhardt, Linda Ludwig, and Joel Stronberg of the Department of Energy; and John Cagnetta (Northeast Utilities), Van Langley (Conoco), Adrian Tiemann (General Electric), and John W. Bartlett (The Analytical Sciences Corporation). We acknowledge their contributions, absolve them of responsibility for the viewpoints expressed, and relieve them of any blame for errors. Elizabeth Peelle of Oak Ridge National Laboratories deserves special thanks for her careful reading of two drafts of the book and for her penetrating insights and helpful comments.

Leroy C. Gould is Professor of Criminology at Florida State University and co-principal investigator with Jan A. J. Stolwijk of a research project entitled "Public perceptions of technological risks." Charles A. Walker is Raymond John Wean Professor and Chairman of the Department of Chemical Engineering at Yale and holds a joint appointment in the Institution for Social and Policy Studies, where he and Dr. Gould co-directed an interdisciplinary social and behavioral science project on energy. Jan A. J. Stolwijk is Professor of Epidemiology and Public Health at Yale. Stanley M. Nealey and John A. Hebert, psychologists at the Battelle Human Affairs Research Centers, have directed several research projects on public attitudes about nuclear energy. Paul Slovic and Baruch Fischhoff are psychologists at Decision Research who have pioneered research in this country on public perceptions of technological risk. Edward J. Woodhouse is Assistant Professor of Political Science in the Division of Science and Technology Studies at Rensselaer Polytechnic Institute and the former Assistant Director of the Yale Institution for Social and Policy Studies.

LEROY C. GOULD

1 THE RADIOACTIVE WASTE MANAGEMENT PROBLEM

When the United States began its nuclear weapons and nuclear power programs, nuclear scientists were fully aware that these industries would generate sizeable quantities of radioactive wastes, including the tailings from uranium mining, effluents from fuel enrichment and fabrication plants the fission and transuranic byproducts of fission reactions, the ordinary metals made radioactive in nuclear reactors by neutron bombardment, and all the materials in all segments of the nuclear fuel cycle that become contaminated through contact with radioactive elements. The need to manage these wastes carefully and ultimately to dispose of them safely, therefore, is not new, although it is daily growing more complicated as the total volume of waste products increases.

Presently only uranium mill tailings and low-level wastes (LLW)—that is, wastes with relatively low levels of radioactivity and with concentrations of plutonium of less than 10nCi/g[1]—are being disposed of on even a semipermanent basis, and some of these may have to be treated further before they can be considered permanently safe (IRG, 1978). Transuranic (TRU) wastes, that is, wastes containing concentrations of transuranic elements[2] above the 10nCi/g threshold, high-level wastes (HLW), and spent fuel assemblies from commercial power reactors are all maintained in temporary storage facili-

1. That is, less than 10 nanocuries (billionths of a curie) per gram. A curie is defined in Chapter 2. The radiation dose to humans represented by 10*nCi/g* is equivalent to the dose from radium in naturally occurring ores in the Colorado plateau.

2. Elements with atomic numbers greater than that of uranium.

ties pending decisions on their final disposition. TRU wastes are being stored in bulk form on retrievable storage pads. High-level wastes, in both liquid and solid forms, are stored in underground tanks at military reservations and at the now abandoned commercial fuel reprocessing plant at West Valley, New York. Spent fuel assemblies from commercial reactors are being stored, intact, in special water-filled storage pools at reactor sites.

FEDERAL COMPLACENCY AND THE GROWTH OF OPPOSITION

In the early days of the U.S. nuclear program, engineers and scientists were more interested in developing new technologies, such as the breeder reactor, than in refining the then-current technologies or in worrying about such seemingly mundane matters as radioactive waste management. Those then knowledgeable about nuclear technology believed that it would be a relatively simple matter, well within the current state of the art, to dispose of low-level and TRU wastes in shallow land burial sites and to process, package, and sequester high-level wastes in deeper geological repositories. With time, however, it became apparent that these early judgments were exceedingly optimistic.

Several years after commercial nuclear power went into production, the U.S. nuclear power industry constructed two plants for reprocessing spent fuel assemblies. Both of these plants have been unsuccessful for various economic, technical, and safety reasons, and both have been closed. There are no immediate prospects of reopening either plant. Construction of a third reprocessing plant was halted by President Carter's decision in 1977 to discontinue commercial fuel reprocessing in the United States until the problem of plutonium proliferation could be resolved. This means that all the spent fuel assemblies from commercial reactors that were not reprocessed before 1977, and very few were, must now remain in storage at commercial reactor sites.

The storage capacity at these reactors, however, is limited.

Some power plants have already reached maximum design capacity. With no reprocessing facilities available, utility companies have few options: they can expand on-site storage capacity, develop temporary spent fuel storage facilities away from present reactors, or shut their reactors down. Several electric power companies have applied for licenses to expand their on-site spent fuel storage capacity, and the Department of Energy has proposed that the federal government build away-from-reactor (AFR) facilities to accept both domestic and foreign spent reactor fuels. It will be several years, however, before such facilities can begin operation.

The complacency that characterized the management of radioactive wastes during the early years of this country's military and civilian nuclear power programs has also been challenged by another development: it has been discovered that existing procedures for radioactive waste management have not always been as safe as they were supposed to be. It has been found, for example, that TRU and low-level wastes have in some cases migrated away from shallow land burial sites.[3] It has also been found that it is not safe to leave uranium mill tailings unattended.[4] And, some of the steel tanks holding liquid high-level defense wastes have leaked. Although none of these breaches has proven to be a major threat to public health or safety, they have in some cases been expensive to repair. They have also served to alert the public to the fact that nuclear engineers, like everyone else, can make mistakes.[5]

Seizing on these accidents and misjudgments, as well as on

3. Since 1970 all TRU wastes, therefore, have been stored on retrievable pads.

4. Such tailings introduce radioactive gases into the atmosphere and are themselves a health hazard if dispersed by winds or water. In the early days of uranium mining some of these tailings were even used to make concrete that went into the construction of homes and sidewalks which have subsequently had to be reconstructed with nonradioactive materials.

5. Although not posing an immediate public health hazard, it has also been discovered that some liquid high-level wastes, notably those at the West Valley reprocessing plant, precipitated radioactive sludges while being neutralized that now present formidable disposal problems (cf. Lester and Rose, 1977).

several mishaps that have occurred in nuclear power plants themselves, antinuclear groups have emerged to demand that the United States abandon its nuclear power program altogether. Nuclear power, these groups argue, is too dangerous, too expensive, and not necessary. Furthermore, they argue, the ultimate safe disposal of high-level wastes is doubtful, and in any event has not yet been demonstrated. Although the nuclear power industry and the federal government have always contended that nuclear power is safe and that radioactive wastes can be disposed of safely and permanently, the credibility of these contentions, particularly concerning the safe disposal of radioactive wastes, has been eroding with time. It is in the interest of the nuclear power industry and the Department of Energy, then, to resolve the "radioactive waste management" issue as soon as possible. It is in the interests of antinuclear forces to defer such a resolution as long as possible. If the issue remains unresolved long enough, these groups contend, there are reasons to believe either that the Nuclear Regulatory Commission will stop granting licenses for new nuclear power plants or that Congress will legislate a moratorium on new nuclear power plant construction, something that several states have already considered and some have in fact made a matter of public law.

ATTEMPTS TO DEVELOP A NATIONAL RADIOACTIVE WASTE MANAGEMENT PLAN

The first public steps toward a rational program for radioactive waste disposal came in 1974 when the Atomic Energy Commission (AEC) issued a draft Generic Environmental Impact Statement (GEIS) covering interim and permanent repositories for transuranic and high-level wastes (AEC, 1974). The Energy Research and Development Administration (ERDA) withdrew this statement in 1975. The Department of Energy (DOE), ERDA's successor, did not reissue the statement until April, 1979 (DOE, 1979b).

In May, 1976, the Energy Resources Council presented what might be called the first comprehensive plan for radioactive waste management (Kuhlman, 1976, p. 18). This program, which would involve six federal agencies, called for a national repository for high-level radioactive wastes by 1985. Although the 1985 date has been abandoned, several features of this early plan are still intact. It is therefore worth noting the major proposals contained in this early document.

> High level nuclear wastes will be processed (extracting and recycling the unspent uranium and plutonium or not according to Nuclear Regulatory Commission decisions still pending) into a stable solid (either a powder or a glasseous solid, with the latter preferred) and encased in steel canisters.
>
> These canisters will be emplaced in excavated repositories in deep (several thousand feet) underground formations of salt, crystalline rock (such as granite), or argilaceous rock (such as shale) that are free from major faults, seismic activity, and ground water intrusion.
>
> The first two repositories, in salt, would be available to receive nuclear waste products by 1985.
>
> Four more repositories, each, like the original two, consisting of a large number of excavated rooms covering an area with a radius of approximately one mile, would be added later and located in other suitable geological formations.
>
> The surface area needed for receiving and handling the nuclear materials would occupy approximately 100 acres with an additional land area of approximately four miles total radius being placed off-limits to drilling and an area immediately over the repository site being leased to the public for restricted purposes.
>
> Waste products would be sequestered in such a way that they could be retrieved quickly in case the repository proved unsatisfactory. Permanent sealing of the repository could occur at a later date if that became desirable.
>
> The selection of repository sites would be made in cooperation with state authorities and after informing and consulting with the concerned public.
>
> Site selection, design, and operation would be subject to licensing and other regulatory authority of the Nuclear Regulatory Commission and would require separate site-specific environmental impact statements.

In anticipation of issuing a revised draft GEIS, ERDA commissioned a complete review of radioactive waste management technology; the results were brought together in a five-volume set of documents, called the Technical Alternatives Document (TAD), in 1977. (This document serves as a major source of data for Chapter 2 of this book.) The TAD document was presented at two public conferences, one in Denver in July, 1976, where technical issues were considered, and one in Chicago in October of the same year, where social, ethical, and policy issues were considered (*Proceedings of Conference on Public and Policy Issues in Nuclear Waste Management,* 1976).

Although the Energy Resources Council proposal could not have assured the states of an official plan without a final GEIS, several other more important developments intervened to lead to this proposal's ultimate demise. The most important was President Carter's decision in 1977 to defer commercial nuclear fuel reprocessing and to consider a wide variety of fuel cycles that were potentially more resistent to proliferation for future U.S. nuclear development. Without commercial fuel reprocessing, various aspects of radioactive waste management would have to be reconsidered and perhaps even redesigned.

In December, 1977, President Carter established a DOE task force to review the government's nuclear waste management program. This task force, supervised by John Deutch, director of the DOE Office of Energy Research, presented its findings in draft form in February, 1978 (DOE, 1978). In March, 1978, President Carter established an Interagency Review Group (IRG) to formulate "recommendations for the establishment of an Administrative policy with respect to long-term management of nuclear wastes and supporting programs to implement this policy" (IRG, 1978, p. 1). Included in the group were representatives from the departments of Energy, State, Interior, Transportation, and Commerce, the National Aeronautics and Space Administration, the Arms Control and Disarmament Agency, the Environmental Protection Agency, the Office of Management and Budget, and the Council on Environmental Quality. This group presented its final report to the president in March, 1979 (DOE, 1979b). Although an extremely important planning document, the IRG report was not a waste management plan. Rather, it was a discussion of the objectives and procedures that such a plan should encompass, technical strategies that could be used, institutional issues that would have to be resolved, and management considerations.

In February, 1980, President Carter issued the following statement to Congress:

> Today I am establishing this Nation's first comprehensive radioactive waste management program. . . . My program is consistent with the broad consensus that has evolved from the efforts of the Interagency Review Group on Radioactive Waste Management (IRG) which I established. . . . My objective is to establish a comprehensive program for the management of *all* types of radioactive wastes.

The Reagan Administration has adopted a radioactive waste management policy that favors reprocessing spent fuels from commercial nuclear power plants and solidifying high-level wastes for emplacement in geologic repositories. Nevertheless, a national plan for radioactive waste management still does not exist, as of this writing, although a preliminary outline for a draft of such a plan has been circulated. Given that this draft plan, once completed, will have to be submitted for public review and comments before a final plan can be prepared, it is fair to say that by the time this book is published, the United States still will not have an official plan for the permanent disposal of radioactive wastes even though it has been in the nuclear armaments and nuclear power business for nearly forty years.

DEVELOPING AN ACCEPTABLE PLAN

Once a draft plan has been issued, the public will be invited to review and comment on what has been proposed. If the plan includes provisions for geologic disposal using conventional mining techniques, as seems almost certain, then an environmental impact statement and plan will have to be drafted for each site and presented to the public for comment. Since many complicated political and social, as well as technical, issues will have to be resolved before any final plan is accepted, however, it is probably not unreasonable to expect that several more years will elapse before a final plan has been approved. After that, still more years will pass before we can begin to dispose of radioactive wastes on anything like a permanent basis. As of the middle of 1982, the schedule for development of the first geologic repository is as follows:

1983	Select three specific sites and begin construction of an exploratory shaft at each.
1985	Complete construction of shafts to repository depth (2000–4000 feet) and begin studies at depth.
1987–89	Construct a Test and Evaluation Facility at one site and place several hundred canisters for a study of waste placement and monitoring.
1988	Select site for first repository and begin design.
1998–2006	Begin operation of first repository.

This schedule must be taken with a grain of salt. Plans have changed before, and it is reasonable to assume that they will change again. Therefore, it might be useful to sketch out some of the other major candidates for a national radioactive waste disposal plan, noting some of the more obvious advantages and disadvantages of each.

One option, suggested by some, would be to quit building nuclear power plants. This is not an unreasonable proposal. It would not, in and of itself, however, solve the problem of radioactive wastes which have been accumulating in temporary storage for more than thirty years and will very likely continue to accumulate for at least that much longer even if we build no new nuclear power generating plants. They will continue to accumulate both because of military programs and because those commercial plants already built have a life expectancy of thirty or more years, during which time they will continue to produce radioactive wastes. To shut down existing plants before their productive years have been realized would be possible, but the investment that went into their construction would be lost and other, non-nuclear plants would have to be built to take their place. It is unlikely, in the absence of some major nuclear disaster, that the United States would be willing to consider scrapping current nuclear plants. It is also unlikely that the United

States will abandon its military nuclear program, in particular its nuclear submarines. Abandoning nuclear power in this country, therefore, would reduce only the size, and to some extent the nature, of the radioactive waste management problem; it would not eliminate it.

Another option, which incidentally is not incompatible with a moratorium on new power plant construction, is to continue storing TRU and high-level wastes and spent fuel assemblies in engineered surface facilities, as is done now. This is not a completely unreasonable idea, and it will be the only option if an acceptable plan for permanent disposal is not developed. Surface storage presents no technical or safety problems that have not already been encountered and dealt with fairly successfully, and for the short term it would at least be cheaper than permanent disposal and would perhaps even be safer (IRG, 1978, pp. 37, 41). Such a program would have the additional advantage of leaving defense wastes at those sites that have already been set aside for them and of leaving commercial wastes at the sites of existing nuclear reactors.[6] Moreover, the longer we wait before committing ourselves to a permanent disposal plan, the more time there will be for scientific investigations of various disposal options and presumably, therefore, the safer our ultimate disposal choice will be.

The major problems with surface storage are the vulnerability of surface facilities, their proximity to the biosphere, and the long periods of time that radioactive wastes will have to be stored. The fact that almost all nuclear waste products to date have been stored in engineered surface facilities and that only relatively minor leaks from these facilities have occurred so far makes it clear that it is possible to store nuclear wastes in this manner, at least for short periods of time. The trouble is that nuclear products remain radioactive, and thus dangerous, for very long periods of time,[7] well beyond the length of time that

6. AFR storage facilities, of course, would entail new storage locations.

7. Chapters 2 and 3 detail this more fully. Suffice it to say here that nuclear waste products could pose a danger to the biosphere for more than 1,000 years and to individuals who might inadvertently come into direct contact with them for more than a quarter of a million years.

any surface storage facilities could be expected to last. This means that nuclear waste products kept in surface facilities for long periods of time would have to be transferred to new facilities at periodic intervals. The old facilities, since they would then be contaminated by radioactive materials, would themselves become waste products. Given these problems, plus the fact that surface facilities are vulnerable to accidents and sabotage, it seems that surface storage is not a viable long-term solution to the management of nuclear waste products. There are no technical or economic reasons, however, why it could not continue to serve as an interim program.

There are, however, as already mentioned, political and institutional reasons why temporary storage probably cannot serve a growing nuclear power industry. At least one Nuclear Regulatory Commissioner has stated publicly that it is unlikely that the NRC will continue to license new commercial power plants beyond the early 1980s unless an acceptable disposal plan has been worked out (Ahern, 1978), and several states already have laws prohibiting new plant construction until the disposal issue is settled. The nuclear power industry, therefore, is less than enthusiastic about temporary storage, which could indirectly cause nuclear plant construction to be discontinued. The military, although by all appearances satisfied with a temporary surface storage option, may also be forced to accept an early permanent disposal plan to ensure against public criticism of its present storage operations.

So long, then, as the federal government continues to view nuclear power as an indispensable part of our national energy plan, so long as antinuclear groups continue to use our lack of a radioactive waste disposal program as an argument against nuclear power, and so long as there are reasons to continue our military nuclear program, pressures to develop a radioactive waste disposal program as quickly as possible will very likely persist.

The draft GEIS issued in 1979 (Department of Energy, 1979a) lists ten potential candidates for ultimate HLW and TRU waste disposal:

1. geologic disposal using conventional mining techniques

2. chemical resynthesis
3. placement in very deep drilled holes
4. placement in a mined cavity in a manner that leads to rock melting
5. island disposal
6. sub-seabed geologic disposal
7. ice-sheet disposal
8. reverse-well disposal
9. partitioning and transmutation
10. ejection into space

Since chemical resynthesis and partitioning and resynthesis are only alternative processing, and not disposal, strategies, there are only eight candidates for permanent disposal. Of these, mined repositories could be made available most quickly, given the present state of technology. It would take several years longer (probably from ten to fifteen) to develop procedures for ice-sheet, seabed, island, reverse-well, or deep-hole disposal. Rock melting and space disposal would take even longer because of the scientific, engineering, and institutional problems that would have to be overcome (Subgroup on Alternative Technology Strategies, 1978).

Chapter 2 describes some of the technical details associated with the more promising of these alternatives, as well as their more obvious advantages and limitations. Suffice it to say here that deep mined repositories is the only alternative that could be implemented soon. Indeed, if repositories were to be available by the early 1990s, they would more than likely have to be in salt formations because salt has been studied more thoroughly than have other geological media. Although rock salt has some definite advantages for repository siting, it also has several disadvantages, and it is not at all clear that salt is ultimately the best medium for mined repositories even if mined repositories are chosen as the best disposal option.

The IRG (1978, pp. 30–31) has outlined four possible technical strategies leading to the disposal of high-level wastes:[8]

8. IRG suggested strategies for disposing of TRU wastes are related to HLW disposal strategies.

Strategy I provides that only mined repositories would be considered and that only geological environments with salt as the emplacement media would be considered for the first several repositories. As a result of past focusing on salt, there is a large volume of information available. In addition, one body of opinion holds that salt is the best, or at least an acceptable, emplacement medium and that suitable sites can be found where salt is the host rock.

Strategy II provides that, for the first few facilities, only mined repositories would be considered. A choice of site for the first repository would be made from among whatever types of environments have been adequately characterized at the time of choice. Because generic understanding of engineering features of a salt repository are most advanced, the first choice is expected to be made from environments based on salt geology. Sites from a wider range of geologic environments would be available for selection somewhat later.

Strategy III provides that, for the first facility, only mined repositories would be considered. However, three to five geological environments possessing a wide variety of emplacement media would be contenders as soon as they had been shown to be technologically sound and economically feasible.

Strategy IV provides that the choice of technical options and, if appropriate, geological environment be made only after information about a number of environments and other technical options has been obtained.

Although not intended as a complete list of all available strategy options, this list is instructive on several points: only strategies I and II offer a chance for an operational repository by the years 1990–2000; strategy IV would take at least ten years longer (Subgroup on Alternative Technology Strategies, 1978). Thus the more time we take to gather scientific data on the several disposal options, the longer we wait to implement a permanent disposal plan.

It should be remembered that the federal government's first tentative plan, proposed in 1976, called for two mined repositories by 1985, an objective that was soon abandoned. Many also question the feasibility of the 1990–2000 dates suggested by the IRG, partly for technical reasons but perhaps more importantly for political and institutional reasons. As the IRG itself notes (1978, pp. 31–32), repository site selection could well run into organizational and political opposition that would interfere with the safety, environmental, and security considerations of an overall repository plan, particularly one that adopted a regional siting approach.[9]

Mindful of past difficulties created by premature site selection, as in Kansas and New Mexico, the DOE has begun a systematic effort to identify the most suitable site for a first repository. Two of the sites being considered are located on government-owned lands. The Basalt Waste Isolation Project (BWIP) is responsible for identifying and exploring a possible site in the basalt underlying the Hanford Reservation in the state of Washington. The Nevada Nuclear Waste Storage Investigation (NNWSI) is responsible for identifying and exploring a possible site in any of several rock types, including tuff (a volcanic ash) underlying the Nevada Nuclear Test Site. A broader search for possible sites in domed or bedded salt and in crystalline rocks, including granite, is being undertaken by the Office of Nuclear Waste Isolation (ONWI), which is also responsible for developing technology that is appropriate for the design and operation of geologic repositories in all media.

Our national radioactive waste management program, then, finds itself somewhere between the frying pan and the fire. The federal government, first through the AEC and more recently through ERDA and DOE, has always maintained that radioactive waste disposal is technologically feasible but, for safety and economic reasons, is best deferred into the future; in the meantime, they rely on temporary storage in engineered surface facilities. Some members of the public, particularly those who are in general opposed to nuclear power, question these assumptions and argue that the production of nuclear power should be terminated until a safe radioactive waste disposal plan has been proven and put into operation. The nuclear power industry, fearing the economic hardships of such an arrangement both for the industry and its customers, seems to be in favor of establishing permanent repositories for radio-

9. Multiple siting would be advantageous for safety reasons and regional siting for reasons of transportation.

active wastes as soon as possible and of restoring the now abandoned program for reprocessing commercial nuclear fuels. The government, being attentive to both industry's concerns for uninterrupted development of electrical power and the public's concerns about safety, must decide between going forward with a disposal plan that could be implemented soon, but might not prove to be the best in the long run, and waiting to explore a variety of options before committing itself to any one. Politically it is a difficult choice.

NUCLEAR POWER IN THE U.S. ENERGY FUTURE

The choice, however, is difficult for more than these political reasons; it is difficult also because choices about the role of nuclear power in U.S. energy development must be made within the complicating context of the "energy crisis." That is, it must be decided not just whether nuclear power, including the management of radioactive wastes, is safe, but whether the U.S. can afford to do without nuclear power even if it is shown to be unsafe. If there are alternatives to nuclear power that are both safer and cheaper, then it is easy to reach the conclusion that the nuclear option can be dispensed with. If there are alternatives that are safer but more expensive, then the decision becomes more difficult. If there are no alternatives at all, except curtailing our energy supplies drastically, then the decision becomes even more difficult.

The difficulty in these decisions comes both from differences in values and uncertainties about energy resources, energy technology, and safety. The purpose of this book is to discuss the safety of radioactive waste management. To decide, however, whether radioactive waste management is safe enough, one must decide in part just how much the United States needs the nuclear option. To decide this it is necessary to review, at least very briefly, the role that energy plays in industrial societies and the nature of the current energy crisis.

The Importance of Energy in Human Social Development

Until the development of agriculture, which began about 12,000 years ago (Cipolla, 1970), human societies were wholly dependent upon but one energy source, the sun, and but two energy converters, fire and their own bodies. Human power, in those days, was responsible for essentially all work; fire provided warmth and some protection from predators and made some foods more edible. The total daily energy budget in preagricultural societies averaged only about 8,000 Calories per person (Brown, 1976, p. 1), 5,000 Calories coming from fire and 3,000 Calories coming from each person.[10] This represents an amount of energy that is released by burning just over one kilogram (2.2 pounds) of coal.

The introduction of draft animals during the agricultural revolution raised per capita energy consumption to perhaps 12,000 Calories per day, or, on a yearly basis, to the equivalent of approximately 600 kilograms of coal per person. The sail, the waterwheel, the windmill, and metal and pottery manufacture, which were developed during the Middle Ages, raised yearly per capita energy consumption to approximately 1,300 kilograms coal equivalent, which is just a little more than three times the per capita energy use of preagricultural peoples. During the 10,000 years of the Agricultural Revolution, per capita energy use little more than tripled.

The Industrial Revolution, which began with the invention of the steam engine approximately 200 years ago, changed this picture dramatically. By the late nineteenth century, yearly world per capita energy consumption rose to 4,000 kilograms (4 metric tons[11]) coal equivalent. In 1970 in the United States, daily per capita energy consumption was approximately 250,000 Calories, or 11 tons coal equivalent per year. This level of energy consumption is 80 times higher than that of preagricultural societies.

The quantities of energy consumed by industrial nations today is truly enormous. In 1970, the most industrialized regions of the world, representing 30 percent of the world's population, consumed 161 million B.t.u. of energy per person per

10. The unit of measure here is the food or large Calorie. The large Calorie equals 1,000 standard calories. One standard calorie is the amount of heat required to raise the temperature of one cubic centimeter of water, at a pressure of one atmosphere, one degree centigrade.

11. The metric ton is equal to approximately 1.1 U.S. tons.

year.[12] Of this amount only about 4.5 million B.t.u. (2.8 percent) was consumed as food, 2.5 million B.t.u. (1.6 percent) as animal feed, and 6.5 million B.t.u. (4 percent) as agricultural wastes or wood (Fisher, 1974, p. 4). The rest, over 90 percent of the energy used in industrialized parts of the world, comes from fossil fuels, nuclear power, or hydropower. In nonindustrial parts of the world the amount of energy from these sources is 50 percent of all energy consumed.[13]

There is little doubt, then, that industrialized societies consume large amounts of energy. There is also little doubt that societies cannot industrialize or remain industrialized without substantial energy budgets, although there is disagreement as to just how big these budgets would need to be if people were as thrifty as possible in their energy use. Some highly industrialized nations, for example West Germany, Sweden, and Japan, get by on per capita energy budgets that are as much as 50 percent below the per capita energy budgets of other highly industrialized nations, such as the United States and Canada (Brookhaven National Laboratory, 1977, p. 25). Nevertheless, if nations that have already industrialized are to maintain their industrial status and if nations that are only now industrializing are to continue their industrial growth, it is clear that they will need large amounts of energy—not as much, perhaps, as the United States and other industrial nations now use, but substantially more than was needed by preindustrial societies.

The "Energy Crisis"

The energy that has fueled industrialization in the past has come predominantly from nonrenewable sources—coal, petroleum, and natural gas. Forty-five percent of current world energy use comes from petroleum, 25 percent from coal, and 19

12. A B.t.u., British thermal unit, is the amount of energy required to heat one pound of water one degree Fahrenheit. One B.t.u. equals approximately 252 standard calories.

13. Total world energy consumption, which reflects both energy consumption per capita and population growth, has grown to the equivalent of seven billion tons of coal per year. This is approximately 1,800 times the total yearly world energy consumption by humans prior to the Agricultural Revolution.

percent from natural gas (Library of Congress, 1975, p. 46). The small remainder comes from such sources as human and animal feed, fuel wood, hydroelectric dams, and nuclear power plants.

At present rates of energy consumption and industrial growth, known reserves of petroleum and natural gas will be all but exhausted in a matter of a few decades (Federal Energy Administration, 1974; Ford Foundation, 1974). Although it is reasonable to expect that additional petroleum and natural gas deposits will be discovered and that existing reservoirs will be exploited more thoroughly than they have been in the past, new exploration and secondary or tertiary recovery in existing oil and gas fields will be expensive and of limited potential given current rates of oil and gas consumption.

The immediate energy problem, then, is that the world is running out of petroleum and natural gas, the dominant fuels in today's industrial economies. The long-run energy problem is that the world will also run out of coal, shale oil, and tar sands. Is there anything that can be done?

In the long run there are but three logically possible, although not necessarily mutually exclusive, answers to this question: (1) revert to an agrarian society that does not need the huge amounts of energy demanded by industrial society; (2) learn how to run industrial societies with considerably less energy; and (3) develop new, inexhaustible energy sources that can continue to fuel industrialization as we know it today. In the short run, of course, the option exists of substituting other fossil fuels—coal, oil shale, or tar sands—for petroleum and natural gas as these two limited fuels run short.

Few people advocate an outright return to agrarian society, although it has been argued that irrespective of our current preferences this may be our ultimate fate (Brown, 1954). A more common position is to advocate movement toward a new form of industrial society that is at the same time less urbanized than present industrial societies and less dependent on transportation and industrial forms that require large amounts of energy (Lovins, 1977; Schumacher, 1973; Woodhouse, 1972). By far the most prevalent position, however, is to assume continued industrial growth within societies that are structured essentially the same as those in present day Europe, Japan, or the United States (ERDA, 1976). Given the dominance of this latter position and the realization that even a restructured industrial society would require more energy than can be provided by agriculture alone, it is important to explore briefly the energy alternatives to petroleum and natural gas.

Other Fossil Fuels

During the early years of the Industrial Revolution, coal was the dominant fuel. Indeed, petroleum and natural gas did not come into a dominant position until the last half century. Given the rather large known reserves of coal in the United States and other parts of the world, what are the prospects for returning to a coal-based energy economy?

The prospects are indeed good. The technologies for mining and burning coal are well known. The trouble is, few home or industrial furnaces are equipped anymore to burn coal directly and none of our current transportation vehicles (automobiles, trains, airplanes, ships, and trucks) can use coal directly as a fuel: most depend on petroleum products. Furthermore, coal is dangerous to mine, more difficult than oil or gas to transport and handle, and contributes to both environmental degradation and air pollution. To return to a coal-based economy, then, would involve both environmental and public health costs.[14] In addition, it would mean that we would either have to replace our present stock of energy converters with those that could use coal directly or convert coal itself to liquids, gases, or electricity that can be used by our present stock of transportation vehicles, industrial machines, and space heating and cooling appliances.

Converting coal into electricity is already common practice in the United States; indeed, it is the major use for coal in the U.S. today. Although no full-scale coal gasification or liquefac-

14. Mining, of course, can be made safer by applying advanced mine-safety techniques, and coal combustion less environmentally polluting by incorporating such things as flue-gas scrubbers. To the extent that this were done, safety and environmental "costs" would be translated into economic costs.

tion plants are operating in the U.S., small-scale gasification plants have operated successfully in the past and both types of plants are today in operation in other countries. The trouble with these plants is that they are large, environmentally disruptive, and expensive. Liquids and gases made from coal in this country would cost at least twice as much per unit of energy as natural gas or liquids made from petroleum do today.

A coal economy would also pose another problem: it might affect world climate adversely. This problem, sometimes called the "greenhouse effect," comes about as follows.

One of the byproducts of burning coal (or any other fossil fuel) is carbon dioxide (CO_2). In and of itself CO_2 is harmless; it is a normal constituent of the atmosphere. As large amounts of fossil fuels are burned, however, atmospheric concentrations of CO_2 increase.[15] This, in turn, some climatologists predict, will raise the average temperature of the atmosphere since CO_2, like window glass, tends to trap solar heat (National Research Council, 1977).

The dynamics of world climate are extremely complicated and, as yet, changes in world temperature have not been linked directly to increased concentrations of CO_2. Nevertheless the potential for such temperature increase is there, and if it were to occur it could have major consequences. It would, first of all, probably shift world climate patterns, making some of today's productive agricultural regions unproductive as it made some unproductive regions more productive. It would also, if the increases were large enough, begin to melt the polar ice caps and thus raise the level of the oceans. This in turn would flood low-lying coastal areas where many of the world's largest cities are now located.

These and many other factors make coal less than desirable as a substitute for petroleum and natural gas. These same factors, moreover, make oil shales and tar sands equally unattractive as major energy sources for the next century. It is under-

15. They also increase through the burning or decay of wood and other organic materials. Indeed, a major contributor to currently observed elevations in atmospheric CO_2 levels has been traced to the deforestation of major world land areas (Woodwell, 1978).

standable that much attention is being given to alternatives to fossil energy.

Nuclear Energy

Although nuclear energy, which must be converted to electricity to be useful, is not as convenient as fossil fuels for some purposes (such as use in transportation vehicles) or as energy efficient, it has the advantages, barring major accidents, of being less harmful to public health (Rowe, 1977), less disruptive to the environment, and inexhaustible.

There are two potential sources of nuclear energy: nuclear fission and nuclear fusion. Nuclear fission, which is described in detail in Chapter 2, involves the release of energy when atoms of uranium or plutonium split into smaller atoms (fission products). Nuclear fusion involves the release of energy when atoms of hydrogen fuse into the larger atoms of helium. In either case, the amounts of energy released are very large indeed. The energy produced by nuclear fusion would be more than adequate to supply the world for billions of years, and the energy produced by nuclear fission could supply the world for thousands (Fisher, 1974).

Controlled nuclear fusion has yet to be demonstrated and, even if it does prove to be technologically and economically feasible, few expect it to contribute significantly to world energy supplies until well into the next century. Present nuclear power plants depend on fission reactions that use uranium as fuel. Known U.S. reserves of uranium used in these plants, assuming that plutonium and unused uranium were not recycled, represent less energy potential than known recoverable U.S. reserves of coal (Fisher, 1974, pp. 31, 34). Breeder reactors based on uranium, such as those now operating in France and the U.S.S.R., would expand the energy potential of this fuel severalfold, and a thorium-based breeder technology would add thorium to the stock of fuels available for nuclear fission. The technology for uranium-based breeder reactors is well advanced, although such reactors have not yet entered commercial power production in this country, and plans for a demonstrated breeder reactor have been curtailed pending a resolution of the problems of nuclear proliferation that this type of reactor poses. Thorium-based breeder reactors are still in the planning stages.

The primary advantage of nuclear power, then, as a substitute for oil and natural gas, is its huge energy potential. Its disadvantages are the inefficiency of current uranium-based, light-water reactors, nuclear proliferation problems associated with uranium-based breeder reactors, and a lack of technological development of thorium-based breeder reactors and hydrogen-based fusion reactors. In addition, nuclear reactors pose the serious threat of nuclear radiation contamination should a major accident befall them, and fission reactors, whether of the conventional or breeder variety, create large quantities of radioactive fission and transuranic byproducts that must somehow be safely disposed of. For these reasons, nuclear power, like coal, is a less than attractive solution to the energy crisis.

Other Energy Sources

Aside from the insignificant energy potential of tides, the only available energy sources besides fossil fuels and nuclear power are geothermal and solar energy. Geothermal energy is heat from the earth's interior; solar energy is heat and light radiated from the sun.

Known recoverable resources of geothermal energy are greater than known recoverable resources of petroleum and natural gas but less than known recoverable resources of coal (Fisher, 1974, p. 35). Like fossil fuels, geothermal energy is also not renewable. Small amounts of geothermal energy are already being used, and research and development is underway to test the feasibility of tapping even larger amounts of this energy resource. Although some people believe that geothermal energy will play a major role in our energy economy in the long run, few believe that it can be developed quickly enough to play much of a role in the short run. Also, geothermal energy is most suitable for low-grade energy uses like space heating or for generating electricity. It could not, therefore, substitute directly for petroleum as a transportation fuel.

Aside from nuclear energy, solar energy is our largest single energy resource. The amount of solar energy reaching the earth each year is approximately 5,300,000 quadrillion B.t.u. (Fisher, 1974, p. 27).[16] This is more than 18,000 times the approximately 290×10^{15} B.t.u. contained in all the coal, petroleum, natural gas, uranium, and hydroelectric power that were consumed in 1978 in the entire world (Library of Congress, 1975, p. 44). Even considering that about one-third of the solar energy reaching the earth is reflected back into space, the world's energy potential from direct solar radiation is enormous.

Humankind, to be sure, has always capitalized on some of this potential. Direct solar radiation plays a major part in keeping us warm and powers the photosynthetic process in plants which provides essentially all our food and a large part of our clothing and building materials. The sun was the source of the energy now contained in all the world's deposits of fossil fuels, in fuel wood, and in wind and running water from which the world today derives over 95 percent of the energy it uses (Library of Congress, 1975, p. 46).

The major disadvantage of solar energy is its diffuseness. Although enormous amounts of solar energy fall each year on the earth's surface, the amount falling on any given spot is small. In New England, for example, the amount of solar energy falling on one square foot of exposed surface in a year is equivalent only to the amount of heat contained in two gallons of fuel oil. Concentrating solar energy, then, into useful quantities at reasonable costs is the major technological challenge facing this energy form. At present, solar heating and cooling are technologically quite feasible but economically marginal.[17] Collectors now available commercially, for example, to capture solar energy in New England cost on the order of $25 per square foot of collector area, installed, and are only about 50

16. It has become fairly common to speak of national or world energy usage in terms of "quads." A quad is a quadrillion or 10^{15} B.t.u.

17. The potential for solar space and water heating and air conditioning, however, is not insignificant. These three energy uses now account for 25 percent of the total United States energy budget (Library of Congress, 1975, p. 53).

percent efficient. This means that it would take a $25 initial investment to claim the equivalent of one gallon of oil per year. As the prices of fuel oil and natural gas go up, however, and as the cost of solar heat collectors goes down, solar heating soon could be a very viable supplement to, if not an outright substitute for, petroleum and natural gas for space and water heating and space cooling.

Eventually, however, we must be able to convert sunlight economically into electricity if the potential of solar energy is to be fully realized. Such conversion is technologically feasible through the use of photovoltaic cells or with large arrays of mirrors that focus sunlight on boilers that create steam. These conversions, however, are not very efficient, nor are they inexpensive. Furthermore, electricity generated in this way would be available only during sunlight hours, necessitating the construction of energy storage facilities before solar-generated electricity could be used widely. The potential for solar energy, then, is great in the long run but limited in the short run.

Transition to New Energy Sources

There is sufficient potential for either nuclear or solar energy to supply the energy demands of growing industrial economies in the long run, but not in the short run, and some doubt whether nuclear power is safe enough for extensive use either now or in the future. Ultimately, then, our energy problem is not one of energy resources, but of safety, technology, and economics as we convert from old energy sources to new ones.

Historically, major energy conversions have taken many years to complete. In 1850, for example, more than 20 percent of the U.S. energy supply came from work animals, less than 10 percent from wind and water, less than 10 percent from coal, and the remainder, 60 percent, from fuel wood (Fisher, 1974, p. 14). By 1910 coal had essentially replaced wood as a source of domestic heat and had substantially reduced U.S. reliance on draft animals as a source of mechanical power. Oil and natural gas combined have materially reduced the use of coal for producing heat and electricity. Hydroelectric power, from about 1890 to 1940, replaced direct wind and water power, and fuel wood today is only an incidental energy source in the United States.

Each of these energy conversions took fifty or more years to complete. Each conversion, moreover, was associated with major changes in lifestyle, transportation, residence patterns, and employment. A major conversion to any new energy source takes time, effort, and capital and results in significant changes in social structure and patterns of living. Current reserves of petroleum and natural gas might be sufficient to sustain society through a period of major conversion to new energy sources if it began immediately and were to take no longer than did the major conversions of the past—that is, fifty or so years. New conversions, however, might not be as rapid as past conversions, for these new conversions will be to sources that are more expensive, less convenient, and in some cases more dangerous than the energy resources we are using now. Furthermore, some potential conversions are not even technologically feasible at this time.

Solar-electric, nuclear fusion, and geothermal power, although very attractive energy sources for the future, are not likely to be developed, for technological and economic reasons, soon enough to substitute completely for oil and natural gas before these two energy resources run out. Solar heat collectors could soon substitute for that portion of our current energy budget that goes into low-grade heating and cooling, although the monetary costs of such conversion would not be small.

Radioactive Waste Management

Radioactive waste management, then, is but a small piece of the total energy picture. It is, however, an important piece as it has the potential for derailing further deployment of nuclear fission technology at just that moment in history when some people feel nuclear power is most needed. Whether or not the reader agrees with this conclusion will depend in part, of course, on just how important he thinks it is for the U.S. to continue its current rate of industrial growth and on whether or not he thinks coal or other fossil energy sources are reasonable short term alternatives to nuclear power. Ultimately, however, the reader must also decide whether he thinks radioactive wastes can be stored and ultimately disposed of safely enough to make nuclear power a viable option in the energy future of the U.S. and the world, both in the short run and in the long run. It is toward this latter decision that the remainder of this book is addressed.

REFERENCES

Ahern, John. 1978. Speech before the Yale Department of Engineering and Applied Science Colloquium, New Haven, Connecticut. December 6.

Atomic Energy Commission (AEC). 1974. *Draft Environmental Impact Statement.* WASH-1539.

Brookhaven National Laboratory. 1977. *Initial Multi-National Study of Future Energy Systems and Impacts of Some Evolving Technologies.* Upton, NY: Brookhaven National Laboratory.

Brown, Harrison. 1954. *The Challenge of Man's Future: An Inquiry concerning the Conditions of Man during the Years that Lie Ahead.* New York: Viking.

———. 1976. Energy in Our Future. In *Annual Energy Review,* ed. Jack M. Hollander and Melvin K. Simmons, vol. 1, pp. 1–36. Palo Alto: Annual Reviews, Inc.

Cipolla, Carlo M. 1970. *The Economic History of World Population.* 5th ed. Harmondsworth, England: Penguin.

Department of Energy (DOE) 1978. *Report of Task Force for Review of Nuclear Waste Management.* Washington, D.C. DOE/ER-0004/D.

———. 1979a. *Draft Environmental Impact Statement: Management of Commercially Generated Radioactive Waste.* DOE/EIS-0046-D

———. 1979b. *Report to the President by the Interagency Review Group on Nuclear Waste Management.* TID-29442.

———. 1981. *Proceedings of the 1981 National Terminal Storage Program Information Meeting.* DOE/NWTS-15

———. 1982. *National Plan for Siting High-Level Radioactive Waste Repositories and Environmental Assessment.* DOE/NWTS-4, DOE/EA-151.

Energy Research and Development Administration (ERDA). 1976. *A National Plan for Energy Research, Development and Demonstration: Cre-*

ating Energy Choices for the Future. Washington, D.C.: U.S. Government Printing Office.

Federal Energy Administration. 1974. *Project Independence.* Washington, D.C.: U.S. Government Printing Office.

Fisher, John C. 1974. *Energy Crisis in Perspective.* New York: Wiley.

Ford Foundation. 1974. *Explaining Energy Choices.* Washington, D.C.: Energy Policy Project of the Ford Foundation.

Interagency Review Group (IRG). 1978. *Report to the President by the Interagency Review Group on Nuclear Waste Management,* draft. Washington, D.C.: National Technical Information Service.

Kuhlman, Carl W. 1976. ERDA Waste Management Program. In *Proceedings of Conference on Public and Policy Issues in Nuclear Waste Management,* Chicago, October 27–29, ed. Harrison Associates, Washington, D.C., pp. 15–27.

Lester, Richard, and Rose, David. 1977. The Nuclear Wastes at West Valley, New York. *Technology Review* 79:20–29.

Library of Congress. 1975. *Energy Facts II.* Prepared for the Subcommittee on Energy Research, Development, and Demonstration of the Committee on Science and Technology, U.S. House of Representatives, 94th Congress. Washington, D.C.: U.S. Government Printing Office.

Lovins, Amory B. 1977. *Soft Energy Paths: Toward a Durable Peace.* Cambridge, MA: Ballinger.

National Research Council. 1977. *Energy and Climate.* Washington, D.C.: National Academy of Sciences.

Proceedings of Conference on Public and Policy Issues in Nuclear Waste Management. 1976. Ed. Harrison Associates, Washington, D.C.

Rowe, William D. 1977. *An Anatomy of Risk.* New York: Wiley.

Schumacher, E. F. 1973. *Small is Beautiful: Economics as if People Mattered.* New York: Harper & Row.

Subgroup on Alternative Technology Strategies. 1978. *Draft Report of the Subgroup on Alternative Technology Strategies of the Interagency Review Group on Nuclear Waste Management.* Washington, D.C.: mimeo.

Woodhouse, Edward J. 1972. Revising the Future of the Third World: An Ecological Perspective on Development. *World Politics* (October):1–33.

Woodwell, George M. 1978. The Carbon Dioxide Question. *Scientific American* 238:34–43.

CHARLES A. WALKER

2 SCIENCE AND TECHNOLOGY OF THE SOURCES AND MANAGEMENT OF RADIOACTIVE WASTES

Discussions of radioactive waste management problems become enlightened when all participants have at least a basic understanding of the sources of radioactive wastes, the materials of which they are composed, and methods that have been proposed for managing them. Such information is essential to understanding the risks posed by these wastes in terms of their effects on humans and other components of the biosphere. Detailed descriptions of radioactive wastes were not readily available until 1976. Since that time the Energy Research and Development Administration (ERDA) and its successor, the Department of Energy, have published detailed descriptions of both commercial and military-related wastes (DOE, 1978a; ERDA, 1976a, 1976b, 1976c, 1977a, 1977b, 1977c). The several relevant publications are thick, ponderous volumes written primarily for an audience with a background in science and engineering. This chapter is a summary of some of the most significant information contained in these volumes.

Before this summary, however, the scientific and engineering background of nuclear fission must be presented briefly in order to define terms and to lend meaning to the basic information on radioactive wastes. Although this material will appear difficult to some at first sight, it is in fact readily understandable to anyone who is willing to accept a fundamental concept: that matter is composed of very small particles called atoms, which in turn are composed of smaller particles called protons, neutrons, and electrons. Beyond that the reader is required to ponder only a few equations for nuclear reactions which, on inspection, turn out to be primarily a convenient method of accounting for protons and neutrons in much the same way that dollars are accounted for in bookkeeping.

NATURAL RADIOACTIVITY

The concept of atoms as small, ultimate, indivisible, eternal particles of which all matter is composed was suggested by the philosophers of ancient Greece. Quantitative meaning was given to the concept beginning in the eighteenth century, when a number of observations in experimental chemistry and physics could be explained only on the basis of the atomic structure of matter. Subsequent determinations of the masses of atoms proved that the particles suggested by the Greek philosophers are indeed small, the mass of the oxygen atom being 2.7×10^{-26} kilograms, for example.[1]

Further developments in physics and chemistry have shown that atoms are neither ultimate nor indivisible nor eternal and that they are in fact composed of three kinds of smaller particles. One of these, the proton, has a mass of 1.6725×10^{-27} kg. and carries a positive electrical charge; the neutron, mass 1.6748×10^{-27} kg., carries no electrical charge; the much smaller electron, mass 9.1091×10^{-31} kg., carries an electrical charge of the same magnitude as that of the proton but of opposite sign. The high degrees of accuracy with which the masses of these very small particles are known reflects the highly sophisticated experimental methods of modern physics.

Protons, neutrons, and electrons are combined in many different ways to make up the atoms of the elements, ranging from the single proton and single electron of the hydrogen atom to and beyond the 92 protons, 143 neutrons, and 92 electrons of a uranium atom. An atom consists of a nucleus containing protons and neutrons, with electrons moving rapidly in a space surrounding the nucleus and held in that space by the electrical attraction between positively charged protons in the

1. In ordinary decimal notation this is 0.000000000000000000000000027 kilograms.

nucleus and negatively charged electrons.[2] A neutral oxygen atom, consisting of a nucleus of 8 protons and 8 neutrons surrounded by 8 electrons, can be depicted in the following way:

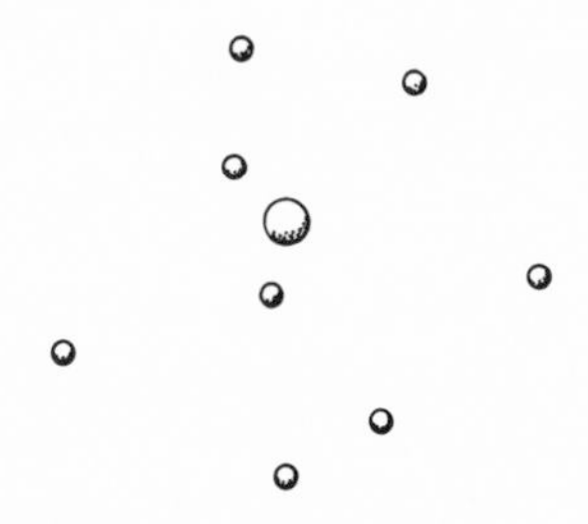

An ordinary chemical reaction, such as the combustion of carbon, is represented by the following equation: $C + O_2 \rightarrow CO_2$. In the course of reaction electrons surrounding the carbon atoms and those surrounding the oxygen atoms are redistributed as though the nuclei were sharing electrons, and it is this sharing of electrons that constitutes chemical bonds between atoms. When this particular chemical reaction occurs heat is evolved, and thus the combustion of carbon, a constituent of coal, can be used to produce heat for the generation of steam or other uses. The quantity of heat evolved by the combustion of carbon is about 9 kilowatt hours per kilogram of carbon.[3]

Reactions involving changes in the nuclei also occur, although they are much less common than ordinary chemical reactions. In fact, until about forty years ago the only known nuclear reactions were the radioactive decays of several naturally occurring heavy elements, including radium and uranium.

2. An astonishing figure from experimental physics is the density of nuclear material, 180,000,000 tons per cubic centimeter. The density of familiar materials is at most a few grams per cubic centimeter. Thus, the presence of electrons surrounding the nucleus results in an effective volume of the atom which is trillions of times the volume of the nucleus.

3. The kilowatt hour (kwh.) is a unit of energy equal to 3,413 British thermal units (B.t.u.). In the form of electrical energy 1 kwh. is enough energy to operate a 100-watt light bulb for 10 hours. In the form of heat 1 kwh., or 3,413 Btu, is enough energy to heat 34.13 pounds (about 4 gallons) of water from 40° F. to 140° F.

More recently, methods of causing certain nuclei to fission, that is, to divide into smaller nuclei, have been discovered and applied in the field of nuclear weapons and the commercial generation of electricity.

A standard method for displaying information about the atomic nucleus is demonstrated in the following:

$$^{235}_{92}U_{143}$$

Here U is the chemical symbol for uranium. The display indicates that the uranium nucleus contains 92 protons and 143 neutrons. The superscript, 235, is simply the sum of these two figures and is thus the total number of nucleons. The chemical nature of an element is determined by the number of protons, and all nuclei with 92 protons represent uranium. The number of neutrons in a uranium atom can be any of several values, including the following:

$$^{233}_{92}U_{141} \quad ^{235}_{92}U_{143} \quad ^{238}_{92}U_{146}$$

Atoms with the same number of protons but with different numbers of neutrons are called isotopes. Thus, the symbols above represent three isotopes of uranium.

Consider the following reaction; an example of natural radioactive decay:

$$^{238}_{92}U_{146} \rightarrow {}^{234}_{90}Th_{144} + {}^{4}_{2}He_{2}.$$

This representation of the reaction is simply a convenient way of keeping track of the nucleons. The symbol $^{4}_{2}He_{2}$ indicates that a particle consisting of 2 protons and 2 neutrons has left the uranium nucleus. Reference to a table of the chemical elements indicates that nuclei containing 2 protons are nuclei of helium. The emission of a helium nucleus leaves 90 protons and 144 neutrons for the other nucleus, and again reference to a table of elements provides the information that this is a nucleus of a thorium atom. The helium nucleus constitutes one form of emitted radioactivity; it is also called an alpha particle.

The thorium nucleus formed in the reaction above also undergoes radioactive decay as follows:

$$^{234}_{90}Th_{144} \rightarrow {}^{234}_{91}Pa_{143} + e.$$

Here the symbol e represents an electron, also called a beta particle, leaving the nucleus. The loss of an electron by a neutron leaves a positively charged particle, or proton. Hence, this equation indicates that the number of protons in the nucleus is increased from 90 to 91, thus forming a nucleus of protactinium, and the number of neutrons in the nucleus is decreased from 144 to 143.

All natural radioactive decay processes result in the emission of alpha particles, beta particles, or a third form of radiation, gamma rays. Gamma rays are electromagnetic in nature, which is to say that they are like light rays. The energy levels of gamma rays are many times greater than those of light rays, however.

Radioactive decay processes occur at rates that are independent of variations in the conditions used by scientists to change the rates of ordinary chemical reactions. Thus, one gram of $^{238}_{92}U_{146}$ decays at the same rate, whether it is in the form of the metal or some compound, whether it is at 0° C. or 1000° C., and whether it is at a pressure of one or many atmospheres. In fact, one gram of $^{238}_{92}U_{146}$ decays at a rate such that one half of a gram will remain after 4,500,000,000 years; that is, this isotope of uranium has a half-life of 4.5×10^9 years. $^{234}_{90}Th_{144}$ decays much more rapidly; it has a half-life of only 24 days.

A second method of expressing the rate of radioactive decay is a unit called the curie (Ci). The curie is defined as 3.7×10^{10} disintegrations per second, or the decay of this number of nuclei per second. One gram of an element such as $^{238}_{92}U_{146}$, which decays very slowly, emits relatively few alpha particles in one second and thus is equivalent to a very small fraction of a curie. One gram of $^{234}_{90}Th_{144}$ decays much more rapidly and is equivalent to many curies. These facts are of central importance in nuclear waste management. Nuclei with long half-lives represent relatively few curies but require attention over a very long time. Nuclei with short half-lives are intense emitters at first, but they decay very rapidly.

NUCLEAR FISSION

A large amount of energy is released when one gram of $^{238}_{92}U_{146}$ undergoes the natural process of radioactive decay, but this energy cannot be used because it is released over such a long time period. Radioactive nuclei with shorter half-lives release energy at much higher rates, but, again, the energy released cannot be put to use readily because the rate of energy release cannot be controlled. In order to use energy from nuclear reactions, it is necessary to develop a way of controlling the rate of energy release so that it can be used to generate steam for use in turbines that generate electricity at a rate equal to demand by customers.

The controlled release of nuclear energy became a possibility about forty years ago when scientists made one of the most significant experimental observations of all time. In studying interactions between neutrons and nuclei, the following unique behavior of $^{235}_{92}U_{143}$ was observed:

$$^{235}_{92}U_{143} \quad + \quad n \quad \rightarrow \quad ^{138}_{56}Ba_{82} \quad + \quad ^{95}_{36}Kr_{59} \quad + \quad 3n$$

Thus is recorded in one line a reaction which is the basis of a nuclear bomb and a basis of commercial nuclear energy, a reaction which some believe to be a boon to humankind and others believe will spell our doom. It is worthwhile to pause and contemplate some basic features of this reaction:

The notation used is again simply a convenient way of keeping track of the nucleons. 92 protons appear on the left side of the equation, and 92 protons (56 + 36) on the right. 144 neutrons (143 + 1) appear on the left, and 144 neutrons (82 + 59 + 3) on the right.

The reaction represents the fission of a large nucleus into two smaller nuclei. In this case the smaller nuclei are those of barium (a metal) and krypton (a gas). Fission also occurs in other ways to yield other smaller nuclei, such as those of selenium, bromine, strontium, and iodine. The smaller nuclei constitute the fission products. They are highly radioactive and create one of the basic problems of nuclear waste management.

When one neutron is absorbed in the fission process three neutrons are released. If one of the neutrons released is subsequently absorbed by another $^{235}_{92}U_{143}$ nucleus, then the reaction continues without further supply of neutrons. The reaction can either become explosive, as in the nuclear bomb, or its rate can be controlled by regulating the flux of neutrons.

The reaction is accompanied by the release of an enormous quantity of energy. The fission of 1 kg. of $^{235}_{92}U_{143}$ yields energy in the amount of 23,000,000 kwh. Compare this with 9 kwh. from the combustion of 1 kg. of coal.

The only fissile isotope occurring naturally in significant quantities is $^{235}_{92}U_{143}$. This isotope is present in natural uranium, which contains 0.7 percent $^{235}_{92}U_{143}$ and 99.3 percent $^{238}_{92}U_{146}$. Since uranium ores currently mined in the United States contain about 0.2 percent total uranium, their content of the fissile isotope is only 0.0014 percent. Thus, more than 70 tons of uranium ore must be processed to yield 1 kg. $^{235}_{92}U_{143}$. Estimates of the total quantity of uranium ore available vary widely, but there is concern that it is probably not large enough to supply significant amounts of nuclear energy for more than a very few decades.

Although no other naturally occurring fissile isotopes are available in reasonable quantities, it is possible to produce fissile isotopes in a nuclear reactor. The most prominent of these is an isotope of plutonium formed by a series of reactions beginning with the absorption of a neutron by $^{238}_{92}U_{146}$:

$$^{238}_{92}U_{146} + n \rightarrow {}^{239}_{92}U_{147}$$

$$^{239}_{92}U_{147} \rightarrow {}^{239}_{93}Np_{146} + e$$

$$^{239}_{93}Np_{146} \rightarrow {}^{239}_{94}Pu_{145} + e$$

The uranium isotope produced in the first reaction rapidly undergoes radioactive decay to produce neptunium, which in turn decays to yield an isotope of plutonium. The $^{239}_{94}Pu_{145}$ so produced is fissile. Since $^{238}_{92}U_{146}$ is present in nuclear fuel, plutonium is produced in significant quantities in nuclear reactors in use today. Recovery of plutonium from spent reactor fuel is thus a way of producing fissile material from the heavier isotope of uranium, which is available in quantities 140 times as great as those of $^{235}_{92}U_{143}$. Fuel reprocessing based on recovery of plutonium and unreacted $^{235}_{92}U_{143}$ will be considered in this chapter, as will developments leading to the breeder reactor based on plutonium.

NUCLEAR REACTORS AND POWER PLANTS

Commercial nuclear reactors can be designed to operate with natural uranium (0.7 percent ^{235}U; 99.3 percent ^{238}U) and are so designed and operated in some countries. In United States practice an enriched fuel containing 3.3 percent ^{235}U and 96.7 percent ^{238}U is preferred because its use allows some simplifications in reactor design. The enriched fuel is produced by a sophisticated separation process called gaseous diffusion, using uranium hexafluoride (UF_6), a gaseous form of uranium. A material balance over a gaseous diffusion plant reveals the following information:

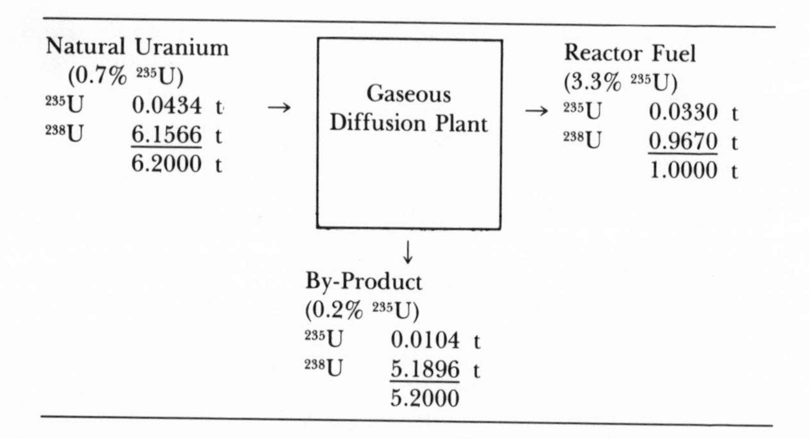

Thus, 6.2 tons of natural uranium is required to produce 1 ton of reactor fuel. The byproduct, containing about 24 percent of the ^{235}U in the original ore, is being stockpiled.

Enriched fuel is manufactured in the form of pellets of uranium oxide. In a reactor the pellets are encased in thin-walled metal tubes, about 1 cm. in diameter and 4 m. in length, sealed at both ends. The metal tubes are arranged so as to be parallel with one another and are supported at the ends by metal plates. A single fuel assembly contains about 300 tubes in parallel and is about 20 cm. square.[4] After these assemblies have served their purpose, they are discharged from a reactor as spent fuel. They constitute by far the most important problem in nuclear waste management.

In the operation of a nuclear reactor the fission reactions occur inside the sealed metal tubes, resulting in heating of the tubes and their contents. Heat is transferred to water flowing past the hot tubes. The flowing water also serves another important purpose in the reactor. Neutrons released by the fission reactions initially have very high velocities. They must be moving at much lower velocities to be absorbed by ^{235}U nuclei. Velocity decreases as neutrons collide repeatedly with the nuclei of substances that do not absorb them. These substances are called moderators. In United States practice, the flowing water serves as a moderator.

Provision must also be made for stopping the fission reactions when a plant is shut down and for controlling the rate of heat release to correspond to a desired rate of steam generation. The rate of fission is most readily controlled by regulation of the neutron flux, and this is accomplished by the presence of substances such as boron which readily absorb neutrons. Control rods containing boron are moved in and out of a reactor to regulate the neutron flux.

A commercial nuclear reactor used in the United States is referred to as a light-water reactor (LWR) to indicate that ordinary water is used as a moderator.[5] A flowsheet for a pressurized-water reactor (PWR), the most common type of LWR in

4. In more familiar units a fuel assembly consists of 300 tubes, each 12 ft. long and 0.4 inch in diameter, arranged in parallel in a square array 8 inches on a side.

5. This distinction is made because heavy water, deuterium oxide, is used as a moderator in reactors used in some other countries.

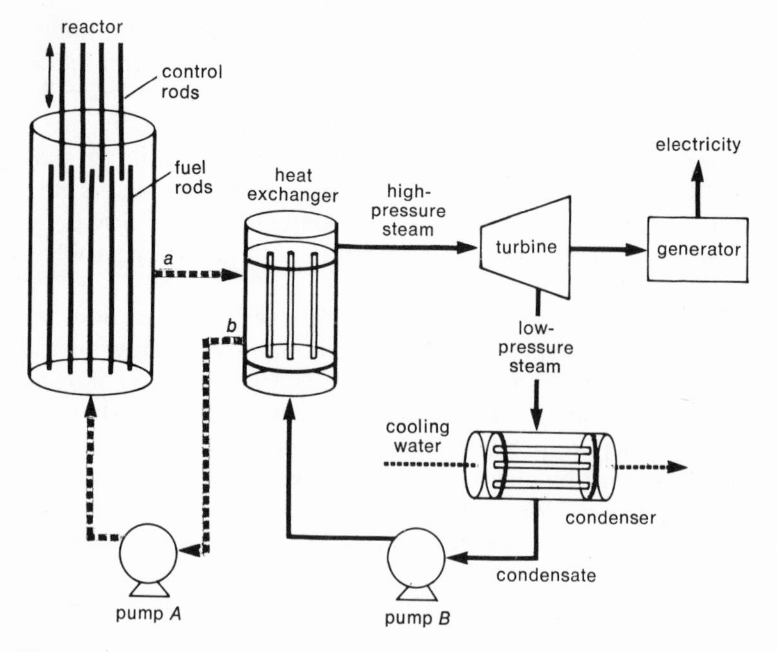

Figure 2.1. Flowsheet of a nuclear power plant

this country, is shown in figure 2.1. In this design pressurized water is circulated over the fuel rods by a pump *A* under sufficient pressure that the water does not vaporize when heated. Heated water leaving the reactor at *a* is circulated through a heat exchanger,[6] where it is cooled and discharged at point *b*, and is recirculated to the reactor. Heat given up by this circulating water in the heat exchanger is absorbed by water flowing inside the tubes at a pressure regulated so that this stream of water evaporates to yield steam at a pressure and temperature suitable for the operation of a steam turbine. The force of expanding steam drives the turbine which in turn drives a gen-

6. A heat exchanger is simply a large cylindrical metal shell capped at both ends and containing many metal tubes fixed into tube sheets. Fluid circulating around the tubes does not mix with fluid circulating inside the tubes but heat flows through the tube walls from the hotter fluid to the colder fluid.

erator to produce electricity. Low-pressure steam from the turbine is condensed in another heat exchanger and recirculated by pump *B*.

In the PWR design, water that circulates around the fuel rods does so in a closed system. This makes it possible to confine any radioactivity in the circulating water owing to suspended or dissolved uranium and fission products picked up by the water from corroded or cracked fuel rods. Concentrations of suspended radioactive materials in the water can be controlled by the use of filters in the closed system, and concentrations of dissolved materials can be controlled by use of a process known as ion exchange. Thus, the level of radioactivity in the circulating water can be controlled at very low values.

The design of a nuclear reactor system would obviously be simplified somewhat if water were allowed to boil in the reactor itself, thus eliminating the need for a heat exchanger for steam generation. This is the basis of the boiling-water reactor (BWR). In this scheme liquid water is pumped into the bottom of a reactor, and steam is removed at the point *a* and fed directly to a turbine system. There is some likelihood of carryover of radioactive materials into the turbine in this design, but the advantages of the simpler system are significant enough to justify the BWR design. Filters and ion exchange can be used on the condensate stream to limit levels of radioactivity.

In discussions of nuclear power plants, it is convenient to know how much uranium is in a reactor and how much uranium is discharged as spent fuel each year. These quantities can be estimated readily if the plant capacity and productivity are known. The capacity of a power plant can be specified in terms of any of several multiples of the watt: the kilowatt (kw.) equals 1,000 (or 10^3) watts; the megawatt (mw.) equals 1,000,000 (or 10^6) watts; and the gigawatt (gw.) equals 1,000,000,000 (or 10^9) watts. The capacity of large nuclear power plants being designed and constructed in the United States today is 1 gw., which can also be expressed as 1000 mw. or 1,000,000 kw.

Energy units must be used to express the productivity of a power plant. As noted above, one such unit is the kilowatt hour. A more convenient unit is the gigawatt year.[7] The output of a 1-gw. plant operated at full capacity for one year is 1 gw. yr., a much more convenient number than its equivalent, 8.76×10^9 kwh. Power plants do not operate at full capacity at all times, of course. Allowance must be made for shutdown for repairs and for periods of low demand. A reasonable estimate of the average productivity of a 1-gw. nuclear power plant over its useful life of about thirty years is 0.62 gw. yr. per year.

In United States practice a 1-gw. nuclear power plant contains about 90 tons of uranium contained in about 200 of the fuel assemblies described above. Since each assembly consists of about 300 fuel rods, the power plant contains nearly 60,000 fuel rods. The amount of spent fuel produced in generating 1 gw. yr. of electricity is about 45 tons. Hence, a 1-gw. plant producing 0.62 gw. yr. per year generates about 28 tons of spent fuel per year.

RADIOACTIVE WASTES FROM CURRENT TECHNOLOGY IN THE UNITED STATES

Radioactive Wastes from Commercial Nuclear Power

The nuclear fuel cycle now in use in the United States consists of the following steps (figure 2.2, solid lines):

Mining of uranium ore containing about 0.2 percent uranium.

Milling to extract uranium from ore by treatment with a chemical reagent. The product of milling is uranium oxide.

7. Estimates of United States installed capacity and productivity by fuel type for 1978 are as follows:

Fuel Type	Installed Capacity (gw.)	%	Productivity (gw. yr.)	%
Fossil fuels	453.7	78.4	187.9	74.7
Nuclear	53.5	9.2	31.5	12.5
Hydroelectric	71.0	12.3	32.1	12.7
Geothermal and other	0.7	0.1	0.3	0.1
Totals	578.9	100.0	251.8	100.0

SOURCE: Department of Energy, Energy Information Administration, Annual Report to Congress, 1978. DOE/EIA–0173/2.

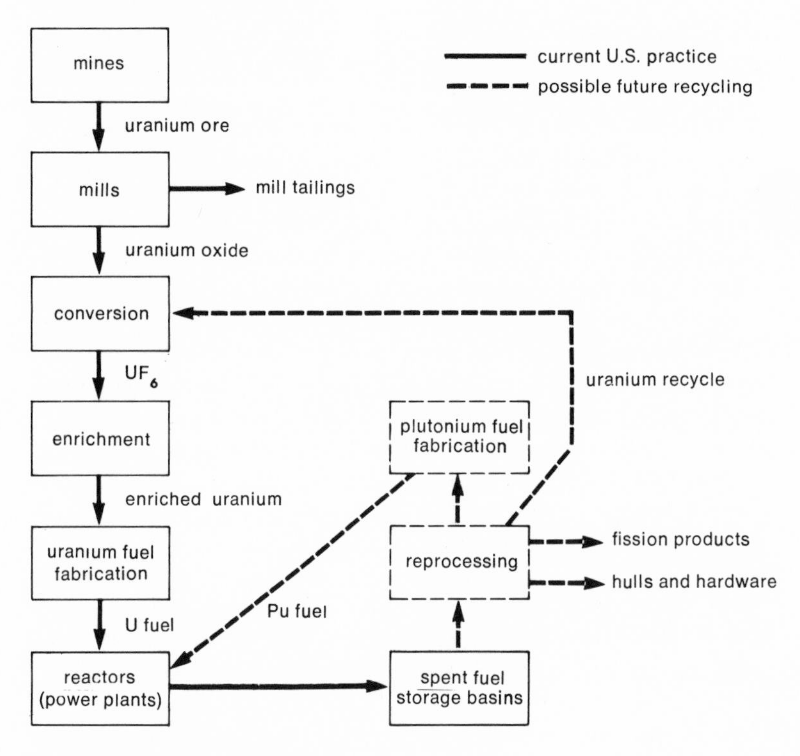

Figure 2.2. Nuclear fuel cycles

Conversion of uranium oxide to uranium hexafluoride.
Enrichment of uranium by gas diffusion.
Fabrication of reactor fuel assemblies.
Reactor operation.
Storage of spent fuel in water-cooled basins at reactor sites.

Each of these steps represents a large-scale industrial operation, and each of them results in the production of radioactive wastes. Waste management systems must be designed and operated to provide for all of the wastes generated over the entire lifetime of each facility, including wastes generated when each facility is decommissioned at the end of a useful life of no more than three or four decades. The operation of nuclear waste management systems extends far beyond the lifetimes of the components of the nuclear fuel cycle because the biosphere must be protected from these wastes for centuries to come.

The first important waste stream in the nuclear fuel cycle is the ore remaining after extraction of uranium, the material referred to as "tailings" from the milling process. Uranium ore contains many radioactive elements other than uranium, including all the radioactive elements in the decay chain of uranium. It has been shown above that the decay chain of ^{238}U begins with the formation of ^{234}Th, which then decays to ^{234}Pa. The decay chain continues through fourteen steps before a stable isotope of lead is formed to stop the process. Extraction of uranium from the ore removes a relatively small part of the radioactivity present, leaving behind tailings that pose a threat to the biosphere. A second important waste stream is generated in the routine operation of reactors and consists of contaminated cloth, paper, ion exchange resins, worn-out equipment, and other items. Tailings, reactor wastes, decommissioning wastes, and other bulky wastes with low levels of radioactivity can be managed by relatively simple methods, such as shallow-earth burial, or, when necessary, placement in geologic repositories. Design and operation of adequate systems will require careful attention to preventing entry of wastes into the biosphere, but there appear to be no significant technical problems to be solved in managing bulky, low-level radioactive wastes.

By far the most important radioactive waste from the current fuel cycle is spent fuel, which contains radioactivity at levels more than a hundredfold greater than the activity of all other wastes combined. The composition and other characteristics of spent fuel are presented in table 2.1.[8]

Table 2.1 contains some of the most important information

8. In this table the abbreviation THM refers to metric tons of heavy metal—that is uranium—charged to a reactor. The term "actinide" refers to an element above actinium in the periodic table, including uranium, plutonium, neptunium, and americium. Note a specifies the quantity of energy obtained from a ton of uranium (25,000 megawatt days), the amount of power obtained from a ton of uranium (35 megawatts), and the initial enrichment.

Table 2.1 Fission Product and Actinide Content of LWR Fuel for Throwaway Cycle[a]

	Mass (kg./THM)[b]	Activity (Ci/THM)[c]						
		90 days	160 days	1 year	10 years	100 years	1000 years	3000 years
1. Total Fission Products	2.6×10^{1}	6.6×10^{6}	4.2×10^{6}	2.1×10^{6}	2.5×10^{5}	2.7×10^{4}	1.8×10^{1}	1.7×10^{1}
2. Actinides								
3. U	9.658×10^{2}	8.3×10^{1}	3.3	3.2	2.7	1.8	2.0	2.1
4. Np	3.3×10^{-1}	6.5	6.5	6.5	6.5	6.5	6.3	5.5
5. Pu	7.8	7.4×10^{4}	7.3×10^{4}	7.1×10^{4}	4.7×10^{4}	2×10^{3}	6.6×10^{2}	5.8×10^{2}
6. Am	7.9×10^{-2}	8.7×10^{1}	1.1×10^{2}	1.7×10^{2}	9.7×10^{2}	2.2×10^{3}	5.3×10^{2}	2.6×10^{1}
7. Cm	8.2×10^{-3}	9.9×10^{3}	7.5×10^{3}	3.5×10^{3}	4×10^{2}	1.5×10^{1}	1×10^{-1}	6×10^{-2}
8. Bk	5×10^{-11}	1.5×10^{-4}	1×10^{-4}	8×10^{-5}				
9. Cf	9×10^{-11}	2.9×10^{-6}	2.8×10^{-6}	2.7×10^{-6}	1.5×10^{-6}	4×10^{-7}		
10. Total Actinides	9.74×10^{2}	8.4×10^{4}	8.1×10^{4}	7.5×10^{4}	4.8×10^{4}	4.2×10^{3}	1.2×10^{3}	6.1×10^{2}
11. Total thermal watts/THM		2.7×10^{4}	1.9×10^{4}	9.2×10^{3}	8.6×10^{2}	1.9×10^{2}	3.8×10^{1}	1.9×10^{1}
12. Total neutrons/second-THM		1.7×10^{8}	1.5×10^{8}	1.1×10^{8}	6.4×10^{7}	1.1×10^{7}	4.2×10^{6}	2.7×10^{6}

a. Basis: 25,000 mw.d./THM, 35 mw./THM, initial ^{235}U enrichment 3.3 percent.

b. Based on material present after 1 year cooling.

c. Curies of activity in fuel cooled for time stated.

Note: Multiply figures in table by 45.6 to obtain quantities per gw.yr. Multiply figures in table by 28.3 to obtain quantities per reactor per year (1-gw. reactor, 62 percent plant factor).

Source: ERDA 76-43, table 2.17.

in the entire literature of nuclear waste management. It should be studied carefully with the following observations in mind:

The table is based on a fuel charge of 1 ton uranium, consisting of 967 kg. ^{238}U and 33 kg. ^{235}U. Although this is not indicated in the table, the text accompanying the table indicates that the spent fuel contains 12 kg. ^{235}U. In simplified form the contents of the spent fuel would then be as follows:

Fission products	26.0 kg.
^{235}U	12.0
^{238}U	953.8
Pu	7.8
Np, Am, etc.	0.4
	1000.0[9]

Fission products (line 1) are highly radioactive 90 days after discharge from a reactor (6,600,000 Ci). They decay rapidly. Radioactivity 10 years after discharge from a reactor (250,000 Ci) is about 3.8 percent of radioactivity after 90 days.

Actinides (line 10) account for only a little over 1 percent of the total radioactivity after 90 days. They decay more slowly and represent most of the total radioactivity after 1,000 years.

Radioactive decay of elements in the spent fuel results in the generation of heat (line 11). The rate of heat generation after 90 days, 27,000 watts, is equivalent to a rate of about 90,000 B.t.u. per hour. When spent fuel is placed in storage, provision must be made to remove this heat by a continuous flow of water.

Although it is not self-evident from this table, the fact is that some of the energy released in commercial power plants results from the fission of plutonium. Consider the simplified table above. The spent fuel contains 26 kg. fission products. The fresh fuel contains 33 kg. ^{235}U, and 12 kg. ^{235}U remains in the spent fuel. Thus 21 kg. ^{235}U fissioned to yield 21 kg. fission products. What is the source of the remaining 5 kg. fission products? Apparently 5 kg. of something else, primarily plutonium, fissioned. Thus 12.8 kg. plutonium was formed in the reactor, 5 kg. fissioned, and 7.8 kg. remained in the spent fuel.

Almost all the spent fuel produced by commercial reactors in

9. Actually, the total mass of spent fuel is about 0.026 kg. less than the mass of uranium charged. This decrease in mass is related to energy by Einstein's famous equation, $E=mc^2$: (energy) = (decrease in mass) (velocity of light)^2.

the United States to date, about 10,000 tons, is currently in storage in water-cooled basins at reactor sites, and this inventory is increasing at a rate of about 1,500 tons per year. The nuclear fuel cycle currently in use could be continued by providing storage in water-cooled basins for periods of ten years or more to allow time for levels of radioactivity and heat generation to decrease. At sufficiently low levels of radioactivity, spent fuel assemblies might be stored in concrete silos with simple air cooling or placed in geologic repositories for permanent disposal.

With table 2.1 in mind, the reader should note that spent fuel contains significant quantities of valuable fissile materials (^{235}U and ^{239}Pu) and a large quantity of ^{238}U, which is potentially valuable in that it can be converted to fissile plutonium. The possibility of reprocessing fuel and recycling ^{235}U and ^{239}Pu (dotted lines in figure 2.2) will be discussed in a later section.

Radioactive Wastes from Military-Related Activities

This chapter is concerned primarily with radioactive wastes from commercial nuclear power plants, but storage and disposal plans for these wastes are closely related to plans for radioactive wastes from military-related operations producing weapons materials such as tritium and plutonium. The plants responsible for weapons production have been in operation for more than thirty years, and a vast amount of information is available from experiences at Savannah River Plant in Aiken, South Carolina, Hanford Reservation at Richland, Washington, and Idaho National Engineering Laboratory at Idaho Falls, Idaho. Radioactive wastes at Savannah River Plant (SRP) are described briefly below as an example of military-related wastes.

Most of the high-level radioactive wastes at SRP are produced as nitric acid solutions of fission products and transuranics. These wastes could have been stored without further treatment, but to do so would have required the use of very large stainless steel tanks. In order to simplify storage problems two steps were taken: the wastes were neutralized with sodium hydroxide to permit storage in ordinary steel tanks and to precipitate some of the dissolved radioactive materials, and water was evaporated from the liquid mixtures to reduce the storage volume required. As a result of these operations, high-level radioactive wastes now in storage at SRP consist of damp crystallized salt, sludge, and solutions with a total volume of about twenty million gallons.

Before any further treatment of these wastes can be carried out it will be necessary to dissolve damp salt in water and use the resulting solution to slurry the sludge. The volume and composition of the reconstituted wastes that would result if wastes accumulated to 1985 were so treated has been calculated, and the results of this calculation provide a convenient basis for describing the SRP liquid wastes (table 2.2).

Several possible methods for storage or disposal of the reconstituted wastes have been considered. These include (1) reconstitution and transfer to new tanks every 50 years, (2) direct

Table 2.2 Radionuclide Content of Reconstituted SRP High-Level Waste (1985)

Radionuclide	Activity (curies)
^{90}Sr	1.3×10^{8}
^{137}Cs	1.3×10^{8}
^{147}Pm	4.6×10^{7}
$^{144}Ce - {}^{144}Pr$	1.1×10^{7}
^{151}Sm	4.2×10^{6}
$^{106}Ru - {}^{106}Rh$	1.8×10^{6}
^{238}Pu	6.0×10^{5}
^{241}Am	6.0×10^{4}
^{244}Cm	6.0×10^{4}
^{239}Pu	2.4×10^{4}

NOTE: Total activity = 320,000,000 curies Total volume of reconstituted wastes = 60,000,000 gallons.

SOURCE: Department of Energy, 1978a, table IV-6

disposal of the liquids to bedrock caverns at SRP, and (3) evaporation of water followed by the conversion of dry solids to glass and disposal in an SRP bedrock cavern or in a federal repository.

Operations at SRP result inevitably in the production of some solid wastes with relatively low radioactivity, such as contaminated equipment and hardware (valves, piping, tanks, pumps, and reactor fittings, for example), contaminated oil and mercury, and contaminated cloth, papers, and plastics. These wastes are placed in shallow-earth burial grounds on the plant site, typically in trenches 20 feet wide and 20 feet deep, well above the mean water table depth of 45 feet. Water transport of radioactive material into the soil is limited to that portion of rainfall that penetrates the soil, a quantity that is limited by surface drainage ditches. Some of the waste is packaged in concrete or other materials, but wastes of very low radioactivity and low TRU content are not packaged. The solid waste area is monitored by a system of wells and boreholes. Data on these wastes are presented in table 2.3.

The problems of storage and disposal of radioactive wastes at Savannah River Plant are large and complex, as are similar problems at Hanford Reservations and Idaho National Engineering Laboratory. Methods for evaporating water from solutions and sludges and for converting the solids to glass or ceramic forms are being studied, but current practice involves primarily continuing tank storage.

Current inventories of radioactive wastes stored at military-

Table 2.3 Solid Radioactive Waste Stored in SRP Burial Ground (through 1975)

Radionuclide	Volume (cubic meters)	Activity (curies)
Actinides	54,500	287,000
Tritium	10,300	2,550,000
Fission products	164,000	58,300
Activation products	22,900	970,000
Totals	252,000	3,860,000

SOURCE: ERDA, 1976b, table 5, p. A-11.

related plants and at commercial power plants can be compared on various bases. On the basis of volume, military-related wastes are much larger than commercial wastes, since the former are in aqueous solutions and the latter are still in the compact form of spent fuel. When the two sources of wastes are compared on the basis of levels of radioactivity, however, they come out nearly equal. Smith (1978) has estimated the total amount of energy generated by military-related operations (plutonium production, nuclear detonations, and naval reactors) and by commercial reactors and has concluded that the total amounts of energy from these two sources were nearly equal at the end of 1975. The quantities of materials fissioned in military-related operations and in commercial reactors up to that time are therefore nearly equal, and radioactive wastes produced are comparable in quantity. Krugmann and von Hippel (1977) have reached a similar conclusion by another type of analysis. At the present time wastes from commercial operations are accumulating much more rapidly than are military-related wastes.

OTHER FUEL CYCLES

The efficiency of utilization of fissile materials in LWRs now in use is relatively low. Fuel cycles that make better use of fissile materials are available, and some of these are described below. As concerns waste management problems, cycles with higher efficiency produce wastes containing smaller quantities of fissile materials. The quantity of fission products per unit of energy obtained is almost independent of the choice of fuel cycle, however, since 1 kg. of *something* must fission to provide each 23 million kilowatt hours.

Reprocessing of LWR Spent Fuel

The fuel cycle that is currently used is referred to as a "throw-away" cycle because the spent fuel contains significant quantities of fissile ^{235}U and ^{239}Pu, both of which could be recovered and recycled. Furthermore, the large quantity of ^{238}U in spent fuel is a potential source of more ^{239}Pu. These facts have led to proposals that spent fuel be reprocessed.

Reprocessing of spent fuel involves basically three steps:

1. Spent fuel assemblies are chopped into small pieces and treated with nitric acid to dissolve the contents of the fuel rods; that is, fission products, uranium oxide, and plutonium oxide. Most of the metal used in the tubes and supporting structure is not dissolved, but remains as a residual waste stream referred to as hulls and hardware (table 2.4).[10]
2. The nitric acid solution from Step 1 is extracted with an organic solvent which dissolves uranium and plutonium compounds. The nitric acid solution remaining after extraction contains most of the fission products and is referred to as high-level liquid waste, HLLW (table 2.5).
3. Uranium and plutonium are recovered from the organic solvent and separated by chemical methods to produce plutonium and a mixture of the two isotopes of uranium.

Reprocessing of spent fuel and recycling of plutonium reduces the amount of uranium ore required by about 30 percent.[11]

This reprocessing technique has been applied to only a few tons of commercial spent fuel in the United States, but it has been used for more than three decades in the production of nuclear weapons based on plutonium as the fissile material. During reprocessing of spent fuel plutonium is produced in a state of high purity after it is separated chemically from ura-

10. Tables 2.4 and 2.5 are based on a dual charge of 700 kg. enriched uranium, 285 kg. natural uranium, and 15 kg. plutonium. In table 2.4 the terms Zircaloy and Inconel refer to metal alloys. TRU is an abbreviation for transuranic elements, which are above uranium in the periodic table and include plutonium and neptunium.

11. Reprocessing of commercial spent fuel was attempted in 1966–71 in a Nuclear Fuels Service, Inc., plant. The plant was closed in 1972, presumably for modifications and enlargement, but NFS has since announced that it does not plan to reopen it. Cleanup costs for this plant are estimated to be as high as $600 million for a plant that cost about $40 million to build (Lester and Rose, 1977). The plant is currently under DOE management and is being used as a demonstration plant for the development of methods for converting liquid wastes to inert, immobilized solids.

Table 2.4 Characteristics of Fuel Assembly Waste (Hulls and Hardware) from Fuel Reprocessing

	Mass	Activity (Ci/THM)			Thermal (w./THM)		
	(kg./THM)	120 days	1 year	10 years	120 days	1 year	10 years
Metal	324	88,000	36,000	3,900	500	190	29
Zircaloy	287						
Stainless steel	29						
Inconel	8						
Radioactive Metals[a]							
Mn	0.6	5,500	320	0.2	45	2.6	1.3
Fe	22	7,100	5,800	520	9.4	7.5	0.7
Co	0.06	9,100	5,200	1,400	140	81	22
Ni	6.8	250	250	240	0.05	0.05	0.05
Sb	0.26	18,000	11,000	1,100	80	50	5
Te	0.02	5,200	4,500	450	5	5	0.5
Fission Products	0.01	2,700	320	120	11	4.5	0.4
Totals/THM	324	91,000	36,000	3,900	500	200	29
TRU elements							
kg./THM	0.01						
Ci/THM		180	150	90			
watts/THM					2	1	0.5
n/sec-THM		2,100,000	1,900,000	1,300,000			

a. These metals are components of the alloys listed above.

NOTE: A 1500 THM/year fuel reprocessing facility generates 486 tons of this waste, or 15.4 tons per gigawatt year.

SOURCE: ERDA-76-43, table 2.15.

Table 2.5 Characteristics of High-Level Liquid Waste from Fuel Processing

	Mass (kg./THM)[a]	Activity (Ci/THM)[b]		Thermal (w./THM)[b]	
		1 year	10 years	1 year	10 years
Fission products	2.18×10^{1}	1.54×10^{6}	2.32×10^{5}	6.69×10^{3}	6.85×10^{2}
Actinides					
U	4.8	4×10^{-2}	3×10^{-2}	2×10^{-4}	4×10^{-4}
Np	2.5×10^{-1}	1.4×10^{2}	1.4×10^{2}	1.9×10^{-1}	1.9×10^{-1}
Pu	1.5×10^{-1}	1.5×10^{3}	1.1×10^{3}	9.7×10^{1}	1.2×10^{1}
Am	8.7×10^{-1}	7.8×10^{2}	7.8×10^{2}	2.3×10^{1}	2.3×10^{1}
Cm	2.4×10^{-1}	3.8×10^{4}	1.7×10^{4}	1.3×10^{3}	6.1×10^{2}
Total actinides	6.25	4.02×10^{4}	1.94×10^{4}	1.38×10^{3}	6.4×10^{2}
Totals	2.82×10^{1}	1.58×10^{6}	2.51×10^{5}	8.07×10^{3}	1.32×10^{3}
Spontaneous fissions (n/second-THM)		3.64×10^{2}	2.49×10^{9}		
Chemicals					
Nitric Acid	4.76×10^{1}				
Others	1.19×10^{1}				

a. Based on masses of various elements after 5 years, 160 days cooling of waste after leaving reactor.
b. After cooling for indicated period following processing at 160 days. One year cooling corresponds to 1.4 years elapsed time from discharge from reactor core; ten years cooling corresponds to 10.4 years.

NOTE: Volume = 378 liters/THM
SOURCE: ERDA 76–43, table 2.9.

nium. Because of its long half-life, 24,400 years, the level of radiation from plutonium is low. It can be stored and moved about in simple containers. These facts have led to serious concern that large-scale reprocessing of spent fuels would make plutonium so easily available that nuclear proliferation would result. Other reprocessing schemes that do not involve the production of plutonium in pure form have been proposed but are not discussed here because of space limitations.

In summary:

The major radioactive waste product of current commercial nuclear reactors in the United States is spent fuel (table 2.1). Spent fuel is now in storage in water-cooled basins at reactor sites.

If reprocessing of fuel were adopted in the United States the major radioactive waste products would be hulls and hardware (table 2.4) and high-level liquid waste (table 2.5).

Several steps of the nuclear fuel cycle result in the production of bulky, low-level radioactive wastes in addition to these major radioactive wastes.

Breeder Reactors

The basic idea of breeder reactors is straightforward. It is based on the following observations:

1. When one ^{235}U nucleus fissions by the equation shown above, three neutrons are emitted. As noted, fission can also occur to yield fission products other than barium and krypton and with other numbers of neutrons. On the average, in commercial reactors about 2.5 neutrons are emitted per nucleus fissioned.

2. Fertile ^{238}U can be converted to fissile ^{239}Pu by the absorption of a neutron.

3. Naturally occurring thorium, $^{232}_{90}Th_{142}$, is also fertile; that is, it yields fissile $^{233}_{92}U_{141}$ after absorption of a neutron and two rapid decay processes.

Of the 2.5 neutrons emitted per fission event exactly one must cause fission of another nucleus in order for a reaction to be self-sustaining. Now suppose that reactor conditions could be designed so that 1 of the remaining 1.5 neutrons is absorbed by ^{238}U or ^{232}Th to yield a fissile nucleus. Under these conditions a reactor would be "breeding" fissile nuclei at the same rate that fissile nuclei are being used. If slightly more than 1, say 1.2, of the remaining neutrons could be made to yield fissile nuclei a reactor could breed fuel faster than it is consumed. This basic idea is interesting because it extends the possible supply of fissile nuclei by utilizing ^{238}U and ^{232}Th, both of which are available in far greater quantities than is ^{235}U.

The engineering problems of breeder reactors are formidable, but France is proceeding with the development of such reactors. United States policy in recent years has not been favorable to such developments, in large part because of the dangers of proliferation of nuclear weapons that could result from the use of plutonium on a large scale.

Because of their higher efficiency in utilization of fissile materials, breeder reactors would produce high-level wastes containing less fissile material than do wastes from LWRs.

Reactors Using Natural Uranium

Development and commercialization in the United States of reactors using enriched fuel and light water as a moderator represent a logical choice in view of the availability of a large enrichment facility built during World War II for weapons production. When Canada began developing nuclear power plant reactors, they logically chose to use natural uranium because Canada had no enrichment facilities. A self-sustaining reaction with natural uranium requires, however, that a more effective moderator be used, and Canada chose heavy water.[12] The resulting CANDU (Canada, Deuterium, Uranium) reactor is in use in several power plants in Canada, and the technology has been exported to other countries.

Space does not permit a detailed comparison of CANDU and light-water reactors. As concerns waste management problems, the CANDU system produces more spent fuel because

12. The nucleus of hydrogen consists of one proton. Deuterium is an isotope of hydrogen, with a nucleus consisting of one proton and one neutron. Deuterium oxide, or heavy water, occurs in very low concentrations in all natural waters and can be recovered in pure form only by complex separation processes. Heavy water is therefore quite expensive.

the fresh fuel contains only 0.7 percent ^{235}U, but, of course, the spent fuel is more dilute in fission products.

Plutonium

Informed choices among fuel cycles can be made only after considering each in terms of all the factors that usually enter into technological choices: resource availability, engineering feasibility, financial analyses, and safety of workers and the public. Comparisons of nuclear fuel cycles must give special attention to a unique factor, the role of plutonium in each cycle, because of the extremely toxic nature of this element and its potential for producing nuclear weapons. For the past several years United States policy has been determined more on the basis of dangers posed by plutonium than on resource, feasibility, financial, and safety considerations.

Much of the concern about plutonium arises from the facts that chemical separation of plutonium from uranium is conceptually simple and that pure plutonium can be handled rather easily because of its low level of radioactivity. The separation could be carried out without appreciable difficulty were it not for the fact that plutonium discharged from light-water reactors is mixed with actinides and highly radioactive fission products. Protection of personnel from this radioactivity, which remains high for many decades after removal of spent fuel from a reactor, makes it necessary that reprocessing be done by complex remote-handling techniques. The complexity of these techniques provides a significant deterrent to reprocessing of spent fuel by unauthorized groups or nations. It is clear, however, that such groups or nations could carry out the separation of enough plutonium to produce nuclear weapons if they were determined to do so.

Aside from attempting to control plutonium by banning reprocessing or devising fuel cycles that produce plutonium only in mixtures with highly radioactive materials, the thorium cycle could be used in conjunction with a uranium cycle. The thorium cycle is dependent primarily on ^{235}U and ^{233}U as fissile materials, and these can be separated from ^{238}U only by complex physical processes, such as gas diffusion, that are far more difficult to design and operate than are facilities for the chemical separation of plutonium from uranium. The thorium cycle is not free of plutonium problems, however, since it must be operated in conjunction with a uranium cycle, and the uranium cycle does result in the production of some plutonium.

TREATMENT, STORAGE, AND DISPOSAL OF RADIOACTIVE WASTES

Wastes from commercial nuclear power plants are troublesome because they emit alpha particles, beta particles, gamma rays, and neutrons, all of which can cause severe damage to biological tissues; they contain radioactive, toxic transuranic elements (TRUs) which do not occur naturally; and they generate heat which can create problems when the wastes are placed in confined spaces. The design and operation of a radioactive waste management system is a complex undertaking. Appropriate systems must provide that wastes are contained in ways that insure against entry into the biosphere over a period of several centuries, that contain radiation, that dissipate heat, and that avoid criticality. These features must be provided at all times and all places in the nuclear fuel cycle, from the time and place of generation of wastes to the time and place of final disposal. The complexity of such an undertaking is indicated by the fact that the accumulation of wastes from commercial power plants and military-related operations in the United States today represents billions of curies of radioactivity in materials currently located in storage basins, tanks, and shallow-earth disposal systems at numerous reactor sites. The design and operation of systems for treatment of these wastes to convert them to suitable forms, transport them to disposal sites, and dispose of them present numerous problems for scientists and engineers. These problems must be studied and solved with full recognition of the fact that the public and its political representatives will insist on participating in making decisions about radioactive waste management systems.

Among the possibilities for disposal sites for radioactive wastes are continental geologic formations, the seabed, ice sheets, and space beyond the earth's atmosphere. Present plans in the United States are to use continental geologic formations for disposal of radioactive wastes, and this possibility is the focus of attention in the following discussion.

The physical and chemical treatment of wastes to convert them to forms most suitable for storage or disposal is discussed at the end of this section. In order to discuss some general issues and some requirements for suitable continental geologic formations before discussing waste treatment, it is sufficient to note here that wastes in the form of compact, inert solids would be most likely to remain where they are placed. Liquid wastes, such as those described in table 2.5, can be converted to inert solids by processes to be discussed later.

Requirements for Systems Based on Continental Geologic Formations

An acceptable nuclear waste management system must provide for containment of the wastes, that is, for insuring that radioactive materials do not enter the biosphere until they have decayed to very low levels of radioactivity over a period of decades or centuries or millenia. Possible routes to the biosphere from storage or disposal sites include burrowing animals and the roots of plants, but these routes are easily avoided by locating repositories at sufficient depths below the surface. Another possible route from repository to the biosphere is carriage of wastes by ground water, a route that is more difficult to avoid. It is possible to locate geologic formations in which ground water flow is not now occurring and has not occurred for many thousands of years. Ground water flow patterns change with time, however, as a result of faulting (creation of breaks in the continuity of a rock formation, caused by shifting of the earth's crust), seismic activity, or glaciation. A third possible route is mining operations by future generations seeking mineral deposits not known to current generations or currently considered to be of no commercial value. Hence, in order to provide containment of wastes over long periods of time, geologic formations must have the following characteristics:

Depth sufficient to prevent access by burrowing animals or the roots of plants

Little or no flow of ground water

Low probability of faulting

Low probability of seismic activity

Stability under conditions of glaciation

Low probability of mining by future generations

Many geologists concerned with these problems believe that suitable geologic formations can be located, but they insist that much more detailed knowledge of formations must be developed before storage or disposal is undertaken. Their caution is understandable since the possibility of placing radioactive wastes in geologic formations raises questions that might not be answered by the results of earlier surveys undertaken for other purposes.

Containment of radioactive wastes must also be provided at all other points of the nuclear fuel cycle, at mines, mills, conversion plants, enrichment plants, fuel fabrication plants, and reactor sites. A nuclear power industry will always involve considerable inventories of wastes stored on or near the surface of the earth. The most highly radioactive of these is spent fuel stored in water-cooled basins for periods of several years while radioactivity decreases to levels low enough to permit safe handling in physical and chemical treatment plants, packaging operations, transportation, and storage or disposal facilities. Bulky low-level wastes such as mill tailings and miscellaneous wastes from power plants will also require careful management to insure that they are not dispersed by mechanisms such as pick-up of dusts by wind or carriage by surface waters.

Radioactive waste management systems must also provide for containment of radioactivity; that is, they must insure that alpha particles, beta particles, gamma rays, and neutrons emitted by wastes do not reach the biosphere in quantities sufficient to cause undue harm to biological organisms. Again a judgment must be made as to how much radiation is "safe." Protection from radiation can be provided by shielding; that is, by surrounding emitters with materials that absorb radiation.

Shielding from alpha particles is a relatively simple matter since these particles, consisting of helium nuclei with two protons and two neutrons, are absorbed by thin layers of paper or metals. The much lighter beta particles, consisting of high-energy electrons, are more penetrating than alpha particles, but they can be absorbed by materials no more than a few millimeters or centimeters in thickness. Gamma radiation, consisting of electromagnetic waves with very high energy levels, is much more difficult to stop. Shielding requirements are rigorous, a 40 mm. thickness of lead being required to reduce gamma activity by a factor of ten, for example. Neutrons penetrate matter readily and can convert matter that absorbs them into radioactive substances.

Shielding is sometimes easily provided, as it is in the case of spent fuel stored in water-cooled basins, where the cooling water provides shielding. It is also provided easily in geologic formations, where the thicknesses of materials surrounding wastes in storage or disposal are very large.[13] The primary concern for containment of radiation is in other parts of radioactive waste management systems. Both spent fuel and bulky low-level wastes from mining, milling, conversion plants, enrichment plants, fuel fabrication facilities, and reactors will require careful shielding at all times between waste generation and storage or disposal. Provision of adequate shielding is particularly difficult when radioactive wastes are being transported. A current design of a transportation system for spent fuel assemblies, for example, consists of an inner cylindrical cask containing spent fuel assemblies immersed in water (ERDA, 1976a). This is surrounded by an annular space containing depleted uranium, which is in turn surrounded by another annular space containing water. The total mass of the empty cask, which must also be provided with equipment for cooling spent fuel, is over 70 tons, and this assembly will accommodate about three tons of spent fuel.

13. Note, however, that some geologists are concerned about possible effects of radiation on the physical and chemical properties of materials surrounding radioactive wastes, and more knowledge about such effects is being sought.

It is interesting to note that shielding from radiation due to radioactive decay of ^{239}Pu is a very simple matter. Because of the long half-life and low specific activity of this alpha emitter, little shielding is required. One unit designed to transport 4.5 kg. of plutonium consists of a steel inner cylinder, an annular space filled with simple insulating material, and a steel outer container. The total volume of the outer container is only about ten gallons. That plutonium can be handled so easily poses serious problems of protection against theft and is one of the reasons that reprocessing of fuel could lead to nuclear proliferation.

A third requirement for radioactive waste management systems is indicated in tables 2.1, 2.4, and 2.5 by data on rates of heat generation due to radioactive decay. The effects of heat generation at all points of the nuclear fuel cycle must be taken into account. Spent fuel elements, for example, generate heat at a rate of 27,000 watts per ton 90 days after removal from a reactor, a rate equivalent to more than 90,000 B.t.u. per hour. If spent fuel were placed in confined, well-insulated spaces, its temperature would rise rapidly, and melting would begin within a few hours. For this reason it is necessary to store spent fuel assemblies in basins cooled by a continuous flow of water.

Such storage can be continued for several years until rates of heat generation and levels of radioactivity have decreased to levels that simplify problems of physical and chemical treatment, transportation, storage, and disposal. The rate at which spent fuel generates heat continues to be significant, however, even after a decade or more of storage in water-cooled basins. As noted above, cooling must be provided in transportation systems. A flow of air from diesel-driven compressors or fans is the cooling agent, but the machinery adds to the volume and weight of the materials, being transported. The heat generated by wastes in storage or disposal in continental geologic formations can be transferred by conduction into surrounding rocks. An effective system of this type requires that the rate of heat conduction be rapid enough to prevent the development of excessively high temperatures in the wastes and in surrounding rocks. Otherwise, thermal stresses might result in fracture or melting of wastes, waste containers, and surrounding rocks and the creation of routes for movement of wastes into the biosphere.

A fourth requirement for radioactive waste management systems arises from the facts that the wastes exhibit a neutron flux, as noted in tables 2.1, 2.4, and 2.5, and that they contain fissile nuclei, ^{235}U and ^{239}Pu. In the presence of water, for example the water in a cooling basin or water leaking into a geologic formation, all of the components required for a chain reaction—neutrons, fissile materials, a moderator—are present. Thus, it is possible that criticality could be achieved in wastes, just as it is achieved in reactors.[14] The result of criticality in even a small part of a waste assembly would be the release of very large quantities of heat, which could overload cooling systems in storage basins or melt rocks in geologic formations. Proper spacing of spent fuel elements or other highly radioactive wastes can eliminate the possibility of criticality under normal conditions. This is not sufficient, however, since criticality must be avoided even if the spacing is changed as a result of such events as a transportation accident or an earthquake. Designers of radioactive waste management systems are therefore required to anticipate such possibilities and to provide for them.

Radioactive waste management systems are thus complex in their design and operation. Containment of wastes, containment of radiation, heat dissipation, and avoidance of criticality can be provided under normal conditions by appropriate design and operation, but much information must be developed before this can be done with confidence. Given a system properly designed and operated, however, instances of unanticipated events and human errors will occur, just as they do in other industrial activities. Thus, waste management systems must be subjected to analyses involving the possibilities of something going wrong in the same way that nuclear reactors are analyzed.

14. The wastes do not contain fissile materials in concentrations high enough for a nuclear bomb to be achieved.

Treatment of Radioactive Wastes

Some desirable characteristics of a system for storage or disposal of radioactive wastes are the following:

The volume of wastes to be stored or disposed of is reduced as far as possible. Solids can be compacted by pressure or by melting and solidifying. Water can be removed from liquid wastes by evaporation. Organic wastes, such as contaminated solvents, cloth, and paper, can be burned to produce small volumes of radioactive ash.

Physical and chemical forms of wastes are selected to limit the probability of migration. Solid chemical compounds with low solubility in water and with little tendency to react with compounds in geologic formations are preferred.

Solid chemical compounds are further limited in the tendency to migrate by immobilization in inert materials. Simple physical incorporation in bitumen is a possibility, as is incorporation in cement mixes. More effective immobilization can be achieved by blending the oxides of radioactive metals into glass or ceramics in the molten state and then cooling to convert to the solid state.

Inert, immbolized solids are encased in corrosion-resistant metals such as stainless steel.

The access of water to inert, immobilized, encased solids is limited by choosing appropriate geologic formations.

In cases where water does penetrate all the barriers listed above and become contaminated with radioactive wastes, the rate of migration of radioactive materials is reduced by chemical interactions with surrounding natural materials such as clay.

This multiple-barrier approach to radioactive waste storage or disposal is necessary because of the very long periods of containment that are required. With these desirable characteristics of waste storage or disposal facilities in mind we now consider the wastes described in tables 2.1, 2.4, and 2.5.

On a basis of 1,000 kg. uranium charged to a reactor, spent fuel assemblies contain about 1,000 kg. fission products and actinides plus 300 kg. structural metals, primarily corrosion-resistant alloys. These spent fuel assemblies are in compact form and could be stored or placed in a disposal site without treatment. Problems of storage or disposal could be simplified, however, if the wastes were separated into various categories.

A process similar to the first steps of recycling could be used for this purpose. Thus, the fuels in the assemblies could be dissolved in nitric acid and subjected to extraction to produce waste streams similar to those described in tables 2.4 and 2.5 and recovered actinides. Hulls and hardware, containing only small amounts of fission products and actinides, could be placed in disposal sites after simple encasement. Liquid wastes containing primarily fission products with very small amounts of actinides could be treated as discussed below. Recovered actinides could be stored for possible future use.

High-level liquid waste (table 2.5) contains fission products and actinides in a nitric acid solution. Evaporation of water from this solution would produce a dry salt cake of metal nitrates. These salts are appreciably soluble in water. In order to obtain less soluble compounds, metal oxides, the evaporation process can be modified by the addition of organic compounds, such as sugar, which react with nitric acid. The solids obtained from such a process could then be immobilized in glass or ceramics.

Gaseous radioactive wastes are emitted in very small quantities during normal operation of a commercial nuclear power plant and in basin storage of spent fuel assemblies. As indicated in the reaction,

$$^{235}_{92}U_{143} \quad + \quad n \quad \rightarrow \quad ^{138}_{56}Ba_{82} \quad + \quad ^{95}_{36}Kr_{59} \quad + \quad 3n,$$

however, a rare gas, krypton, is one of the many products of nuclear fission. This gas is emitted when spent fuel is dissolved in nitric acid, the first step of a recycling process. Quantities of this gas and another rare gas that is emitted, xenon, are estimated as follows in ERDA-76-43:

Gas	kg./THM	Ci/THM
Kr	0.25	6,860
Xe	4.06	127

This amounts to approximately 200,000 Ci per year from a reactor discharging 28 tons per year spent fuel.

These gases pose a unique problem in radioactive waste management since they cannot be converted to solid compounds. They could be dispersed to the atmosphere from tall stacks, a practice now followed in military-related plants. If such discharge were not permitted, it would be necessary to recover the rare gases by a process similar to widely used industrial processes for separation of air into oxygen and nitrogen by low-temperature liquefaction and distillation. Recovered krypton and xenon could then be compressed and stored in thick-walled steel cylinders. Safe storage of such compressed gases for periods of at least several decades would require careful design and maintenance.

Alternatives to Storage or Disposal in Continental Geologic Formations

Several environments other than continental geologic formations could be used for disposal of radioactive wastes, including the seabeds, ice sheets, and extraterrestrial space. Retrieval of wastes from these environments would be difficult; thus, they are not considered to be suitable for storage.

Placement of radioactive wastes in the seabeds poses a risk that these materials could be dispersed through the oceans and enter biological chains through the medium of aquatic life. This risk can be minimized by placing wastes in the consolidated sediments of the seabeds, which lie below a layer of more recent, unconsolidated sediments. In fact, some of the consolidated sediments at the bottom of the ocean are known to have been stable for millions of years and constitute geologic formations as "safe" for disposal as are continental formations. Two factors work against the use of the seabeds for radioactive waste disposal, however. One is the fact that the technology for placement of wastes under deep water would be more complicated and probably more expensive than placement in land masses. A second limitation is imposed by the difficulty of achieving agreement among nations on the use of the seabeds for radioactive waste disposal, particularly when the possibility of accidents in waste-placement technology is considered.

The ice sheets appear at first glance to be quite suitable

places for radioactive waste disposal because very simple placement techniques can be devised. It would be possible, for example, to place containerized, immobilized wastes on top of the ice sheets, where heat generation due to radioactivity would melt the ice, allowing the wastes to sink to underlying rock. Physicists and geologists who study the ice sheets warn, however, that their ability to predict the consequences of such a disposal method are limited, and they advise against placement of wastes in the ice sheets.

Disposal of radioactive wastes by ejection to extraterrestrial space would be a costly process, even if the wastes so ejected were limited to the separated fission products and transuranics. This cost would be increased still further by the necessity for packaging wastes to survive accidents during launching. These considerations alone probably preclude the disposal of radioactive wastes in extraterrestrial space.

Simplification of storage and disposal problems might be realized someday by transmutation; that is, by converting the radioactive nuclei of the wastes to nonradioactive nuclei or to nuclei with shorter half-lives. Thus, plutonium, with a half-life of over 24,000 years, can be converted to products with half-lives of a few hundred years or less by fission resulting from collisions with neutrons. Other radioactive nuclei from fission reactors are not fissile, however, and transmutation of the bulk of radioactive wastes would have to be achieved by processes not yet known. Hence, transmutation does not appear to be a solution for the next several decades, if ever.

A Natural Experiment in Radioactive Waste Isolation

A self-sustained nuclear fission process becomes possible when conditions are established such that, on the average, at least one of the 2–3 neutrons emitted by the fission of one ^{235}U nucleus is absorbed by another ^{235}U nucleus and causes fission. Whether or not such a reaction will occur depends in a complex way on the total amount of uranium present, the concentration of uranium, the $^{235}U/^{238}U$ ratio, the presence of a moderator, and the concentrations of elements such as boron that absorb neutrons. Man learned to provide and maintain appropriate conditions about forty years ago.

Nearly thirty years ago a question was raised as to whether conditions for a fission reactor might ever have occurred in nature. An answer to this question began to develop in 1972 in a laboratory in France when analysis of some particular samples of uranium ore indicated that the uranium in these ores contained slightly less ^{235}U than is characteristic of most ores (0.7171 percent as compared with 0.7202 percent) (Cowan, 1976). One of the possible explanations of this lower concentration of ^{235}U is that, sometime in the past, processes of nature combined to establish conditions necessary for a self-sustaining reactor. Tracing the ore back to its source in an open-pit mine in Oklo in the Gabon Republic of West Africa, scientists analyzed materials containing uranium deficient in ^{235}U and found the elements that would be expected to result from fission. They estimated that a natural reactor was active for 150,000 years about 1.8 billion years ago.

This discovery affords a natural experiment in radioactive waste management in that it is possible to ask and answer questions about the migration of fission products, unreacted uranium, and transuranics in this particular body of ore. The answers are encouraging in indicating that in this case migration was extremely slow and geologic barriers appear to have provided a reasonable basis for long-term management of radioactive wastes.

COST ESTIMATES

Estimates of the costs of storage and disposal of radioactive wastes are difficult to make because of many uncertainties about the technologies to be used. Such estimates must be attempted in spite of the uncertainties because of their direct bearing on the future of commercial nuclear power. Furthermore, the estimates must include consideration of the relative values of the benefits (reductions in exposure of living organisms to the effects of radiation) and the costs of achieving various levels of reduction in radiation exposures. The cost figures discussed below must be viewed as approximations only. The history of cost estimating for new technologies suggests that such estimates are likely to be appreciably lower than actual costs.

Costs of storage and disposal of commercial spent fuel in the throwaway mode are of obvious importance in determining how much utilities should be charged for services to be provided through the federal government. A preliminary estimate (DOE, 1978b) by the Department of Energy is $232 per kilogram (based on 1978 dollars) for storage ($104), transportation ($26), encapsulation ($28), geologic repository ($42), research and development ($26), and government overhead ($6). On this basis, the cost of storage and disposal of the existing stock of approximately 10,000 tons of spent fuel would be $2.3 billion; the annual cost for a current production rate of 1,500 tons per year would be $0.4 billion. A more recent estimate includes $2.9 billion for research and development and between $11 billion and $18 billion for total waste management costs, assuming that current reactors continue to operate for normal lifetimes and using 1990 as the date for a first repository (DOE, 1980). The cost per unit of electrical energy produced would be about one mill per kilowatt hour, a small fraction of the total cost of electrical energy.

Utilities should also assume responsibility for the eventual decontamination and decommissioning (D & D) of commercial nuclear power facilities. Among the technologies available for D & D of nuclear power plants are dismantlement (total removal of the facility from the site to waste disposal systems), entombment (sealing of a reactor system with concrete or steel after liquid waste, fuel, and surface contamination have been removed), mothballing (placing a facility in protective storage after simply removing fuel and radioactive waste), and entombment or mothballing followed by dismantlement after 50–150 years. Here again the costs are uncertain since limited experience with this stage of the nuclear fuel cycle is available. D & D problems are reviewed in a report by the General Accounting Office (1977) which includes estimates ranging from

$3 million (1975 dollars) for mothballing a power plant to $40 million for dismantlement. These are costs that will be paid several decades in the future, and they create serious problems of intergenerational equity.

Other components of the commercial nuclear fuel cycle also generate radioactive wastes and will eventually require D & D. A case in point is the Nuclear Fuel Services plant at West Valley, New York, which was designed to reprocess spent fuel from commercial reactors. Since this plant is not scheduled to reopen, the problem of D & D is posed now. Estimates of the cost of D & D of this plant, which originally cost about $40 million, range from $90 million to $600 million (Lester and Rose, 1977).

Costs of waste management operations at military-related plants have also been estimated. For the Savannah River Plant, for example, budgetary costs of waste management alternatives have been estimated to be $240 million for continued tank farm operation and $2.7 billion for conversion to glass and shipment to an offsite repository (ERDA, 1977a). These costs, in 1976 dollars, refer only to disposing of existing wastes. The costs of D & D for military-related plants are illustrated by an estimate of $4 billion for D & D of the Hanford facilities (ERDA, 1977c).

The total bill for disposing of the existing wastes and for D & D of all existing facilities, commercial and military-related, is probably on the order of $50 billion.

TECHNICAL ISSUES IN RADIOACTIVE WASTE MANAGEMENT

Commercial Spent Fuel

The following discussion is limited to consideration of alternatives for managing the existing stock of commercial spent fuel in the United States, which now amounts to about 10,000 tons, and the spent fuel currently being produced in light-water reactors at a rate of about 1,500 tons per year. Only the fuel cycles and management schemes described in this chapter are considered. Other fuel cycles are possible, and other methods of waste management will probably be developed in the future, but these are likely to change the basic problems only in degree, not in kind.

Basically, three methods are available for management of spent fuel assemblies: spent fuel assemblies can be subjected to ultimate disposal in a throwaway mode; they can be managed in a stowaway mode awaiting decisions about reprocessing; or they can be reprocessed to recover fissile and fertile materials. A choice among these three modes is one of the most important decisions about radioactive waste management to be made in the near term. The reprocessing mode was banned during the Carter administration, but it is favored by the Reagan administration. In fact, however, we are currently operating by default in a stowaway mode, and spent fuel is simply accumulating in water-cooled basins at reactor sites.

The throwaway mode, the placement of spent fuel assemblies in backfilled and sealed geologic formations, appears attractive because of its simplicity and seeming finality. It appears attractive also because wastes so placed would reduce worries about safety, vulnerability to theft or sabotage, and proliferation. Furthermore, surveillance requirements would be minimal. Disadvantages of the throwaway mode include the loss of large quantities of fissile and fertile materials, the necessity for planning on a time scale determined by the elements with the longest half-lives, and uncertainties about the effects of heat release and radioactivity on geologic formations.

If the throwaway mode were adopted decisions would have to be made about pretreatment of spent fuel assemblies, generic site selection, and specific site selection. Pretreatment would probably be limited to simple mechanical procedures: spent fuel assemblies might be disassembled to separate hardware from fuel, mechanical compaction might be used to meet volume requirements of the repositories, and cladding in corrosion-resistant materials might be used to bar migration of the wastes.

Generic site selection refers to the development of criteria for suitable disposal sites. Among the site characteristics to be considered are the type of geologic formation (salt, basalt, etc.), the nature of materials surrounding host rocks, groundwater flows, seismicity, faulting, and glaciation. Choices must also be made about the number of sites to be used, methods for handling and placing wastes, transportation, and surveillance requirements.

Once suitable criteria for disposal sites are developed the process of selecting specific sites begins. This process involves interactions among government agencies at the federal, state, and local levels. It also involves interaction between government agencies and members of the public and special-interest groups.

Implementation of the throwaway mode is thus a complex process, even though this is the simplest mode of radioactive waste management. Adoption of this mode would mean, however, that significant quantities of fissile and fertile materials would be lost. Spent fuel contains a higher concentration of ^{235}U than does natural uranium. It also contains significant quantities of fissile ^{239}Pu and a very large quantity of ^{238}U which can be converted to ^{239}Pu by neutron absorption. In fact, present practice in the United States provides for utilizing only about half of the 7 kg. of ^{235}U present in a ton of natural uranium. It also provides for utilizing a very small part of the 993 kg. of ^{238}U in a ton of natural uranium since some ^{239}Pu is both formed and fissioned in reactors. Overall, the amount of energy available from a ton of natural uranium is more than two hundred times greater than the energy used by current light-water reactors. Utilization of all this energy would require the development of breeder reactors, but significant gains in efficiency can be made by the simpler reprocessing techniques described in this chapter.

Given the considerable value of spent fuel it is reasonable to think of managing it in a stowaway mode. As noted above, current United States practice follows this mode by default. Implementation of the stowaway mode would require a set of decisions similar to those of the throwaway mode; that is, decisions would have to be made about pretreatment of spent fuel, generic site selection, and specific site selection. In this case generic site selection would involve the design and development of storage methods and structures. After storage in water-cooled basins for a period of several years spent fuel assemblies might be stored in air-cooled concrete vaults located on the surface of the earth or in underground caverns. This mode would provide for recovery of fissile and fertile materials in the future, when and if reprocessing and breeder reactors are developed. It does not have the seeming finality of the throwaway mode, but it provides time for further studies of geologic disposal. Disadvantages of the stowaway mode include more severe surveillance and monitoring requirements and greater risks that wastes in storage at or near the surface would enter the biosphere.

The method described earlier for separating spent fuel into uranium, plutonium, hulls and hardware, and fission products could be used as a basis for simplifying storage and disposal problems. Uranium and plutonium, representing 97 percent of the total spent fuel, are low in radioactivity and heat generation rates and can be stored in simple, shielded structures. Hulls and hardware are relatively low in radioactivity and heat generation rates (see table 2.4) and could be placed in geologic formations with little pretreatment. The most significant waste disposal problem in the reprocessing mode is posed by the nitric acid solution of fission products since these materials are highly radioactive and generate heat at significant rates. On the other hand, this solution contains only 26 kg. of fission products for each ton of uranium charged to a reactor. The fission products can be recovered as solids by evaporation of water, and they can be converted to glass or ceramic forms for disposal. The technology for recovery and conversion of fission products is not fully developed, but the basic steps have been demonstrated. Further development of such technology is essential for processing the existing millions of gallons of high-level liquid wastes from military-related operations. The results of such development could be applied to liquid wastes created in the reprocessing of commercial spent fuel.

The reprocessing mode thus saves fissile and fertile materials and results in reductions in the magnitude of waste disposal problems. Disadvantages of this mode include the fact that a large new industry would be created, and the existence of this industry would lead to increased risks of contamination of the biosphere and, ultimately, to problems of decontaminating and decommissioning still more industrial plants. Surveillance and monitoring requirements would be particularly severe in this mode.

If the reprocessing mode were adopted, decisions about waste forms for disposal and about generic and specific disposal site selection would again be required. It would also be necessary to undertake generic and specific site selection for reprocessing plants. This process would be complicated in view of the large quantities of radioactive materials shipped into and out of the plants and the risks of increased levels of radioactivity in areas surrounding the plants. Siting problems for reprocessing plants would also be made more difficult by the dismal record of the Nuclear Fuel Service plant at West Valley, New York.

High-Level Waste from Military-Related Operations

Large quantities of high-level wastes have accumulated during more than three decades of nuclear weapons production in the United States. These wastes are primarily in the form of aqueous solutions, sludges, and salt cake and are currently in storage in steel tanks at sites in South Carolina, Washington, and Idaho. Methods currently being studied for disposal of these wastes include removal from storage tanks, evaporation and drying to recover solids, conversion to glass or ceramic forms, cladding in corrosion-resistant materials, and placement in continental geologic formations.

Decisions to be made about these wastes are similar to those described above for commercial spent fuel with operations in a reprocessing mode.

Low-Level Wastes

Commercial nuclear power and military-related operations result in the generation of low-level wastes in the forms of discarded equipment, ion-exchange resins, construction materials, clothing, and other material. As noted earlier, solid wastes in shallow-earth burial sites at Savannah River Plant alone amount to 250,000 cubic meters (9,000,000 cubic feet) with a total radioactivity of four million curies. Low-level wastes from military-related operations of other sites in the United States amount to 40,000,000 cubic feet, and 16,000,000 cubic feet of commercial low-level wastes are buried at several sites. Some perspective on the magnitude of these wastes is provided by noting that the total volume is equivalent to that of a cube 400 feet on a side. It is also equivalent to the volume of a 10-foot depth of waste distributed over 140 acres.

Plans for future disposal of solid low-level wastes include continuation of shallow-earth burial for wastes containing concentrations of transuranic elements (TRUs) below 10 nanocuries per gram (1 curie per 100 tons). Wastes with higher concentrations of TRUs were also placed in shallow-earth burial prior to 1975, but plans for the future are that such wastes will be placed in geologic repositories.

Sites for burial of low-level, low-TRU commercial wastes exist now at Barnwell, South Carolina; Beatty, Nevada; Hanford, Washington; Maxey Flats, Kentucky; Sheffield, Michigan; and West Valley, New York, but the sites in Kentucky and New York are now closed. Sites for burial of low-level, low-TRU military-related wastes are at Hanford, Washington; Idaho Falls, Idaho; Los Alamos, New Mexico; Oak Ridge National Laboratories, Tennessee; Savannah River, South Carolina; Nevada Test Site; and Sandia Laboratories, New Mexico.

Commitments of sites to shallow-earth burial of low-level, low-TRU solid wastes have thus been made. These commitments could be changed for the disposal of future wastes, but a reversal of the decision about wastes already buried could be achieved only at the expense of exhuming such wastes and transporting them to new sites.

Decontamination and Decommissioning (D & D)

At the end of useful lives estimated to be 30–40 years, commercial nuclear power plants will be transformed from useful plants to a unique type of junk. The materials present in retired plants will include large volumes of concrete, metals, and other material, ranging in radioactivity from high levels for materials near the reactor cores to very low levels for exterior walls.

Decisions are yet to be made about the treatment of retired reactors. They could be entombed in place after dismantling and removing some of the most highly radioactive materials, left to decay for periods of decades or centuries, and ultimately dismantled. At the other extreme they could be dismantled soon after retirement, separated into the then-current classes of radioactive wastes (high-level; low-level, low-TRU; low-level, high TRU), and managed as are similar wastes from other parts of the nuclear fuel cycle.

Decontamination and dismantling of commercial nuclear power plants can be expected to begin in the 1990s in the normal course of events. It is possible, however, that an accident such as the one at Three Mile Island will disable a reactor so severely that D & D will be the most economical way out.

Decisions about D & D of commercial reactors, whenever these processes come about, will be difficult indeed, involving questions about intrusion, surveillance, and intergenerational transfers of responsibility. D & D activities are not limited to commercial reactors, of course. Every component of the commercial nuclear fuel cycle (mines, mills, conversion plants, enrichment plants, fuel fabrication plants, reprocessing plants, and shipping casks) and every component of military-related plants will be ultimately a D & D problem.

WASTE MANAGEMENT PROBLEMS OF OTHER ENERGY TECHNOLOGIES

This book is about radioactive wastes from nuclear power plants and military-related operations, and these wastes are depicted here as posing very serious management problems. Wastes from other energy technologies also pose serious problems, and choices among various sources of energy must include comparative analyses of waste management problems and environmental impacts as well as cost considerations.

One possibility for energy for the future is an expansion in the use of coal, which can be utilized by direct combustion or converted to liquid and gaseous fuels. Significant expansion in coal use will have numerous impacts, beginning with a large increase in the mining work force and vast social changes in areas where coal is available. Environmental impact on land, air, and water will be severe. For example, a coal-fired power plant operating at a 62 percent capacity factor requires three million tons of bituminous coal per year (the output of a mining work force of 1,500 workers in underground mines), and it results in the generation of about 300,000 tons of ash and 180,000 tons of sulfur dioxide per year. These wastes are more than a thousand times larger in mass than the wastes from a nuclear power plant producing the same quantity of electricity.

Even that favorite of many people, solar energy, can be utilized effectively only by the use of enormous quantities of steel, aluminum, copper, concrete, plastics, and lumber in solar collectors and photovoltaic cells. The production of each of these materials results in the generation of wastes and in environmental impact.

Hence in thinking about radioactive waste problems it is well to bear in mind the fact that all energy technologies result in generation of wastes. The wastes from nuclear power plants are far smaller in mass than those of other energy technologies, but they pose management problems of a kind that we have not encountered before.

COMMENTS

The preceding discussion of the nature and magnitude of radioactive waste management problems is intended to be informational. Whether or not it is also alarming depends in part on one's faith in the ability of technologists to solve the problems and in part on one's faith in the ability of social and political institutions to recognize appropriate programs and to administer them. The technological problems described here can be solved, and, in fact, many of them are solved in principle by the results of the considerable amount of research and development on radioactive waste management that has been carried out in the past few years. Further research and development will be required to arrive at optimum methods, of course, but at present we have no choice other than to develop acceptable methods to store or dispose of current inventories of commercial and military-related wastes.

If we can develop methods of managing radioactive wastes with low risks to workers and to the public, as appears likely, and if we can manage these wastes at acceptable financial costs, which is less certain, there remain serious problems to be solved by social and political institutions. The wastes must be managed for a period of at least several centuries. Control of the wastes must be continuous for these several centuries, effective in times of economic difficulties as well as in times of prosperity, and effective in times of civil disturbance as well as in times of peace.

REFERENCES

Cowan, George A. 1976. A Natural Fission Reactor. *Scientific American* 235:36–47.

Department of Energy (DOE). 1978a. *Draft Environmental Impact Statement: Long-Term Management of Defense High-Level Radioactive Wastes, Savannah River Plant, Aiken, South Carolina.* Washington, D.C. DOE/EIS-0023-D.

———. 1978b. *Preliminary Estimates of the Charge for Spent-Fuel Storage and Disposal Services.* DOE/ET-0055.

———. 1980. *Final Environmental Impact Statement. Management of Commercially Generated Radioactive Waste.* DOE/EIS-0046 F.

Energy Research and Development Administration (ERDA). 1976a. *Alternatives for Managing Wastes from Reactors and Post-Fission Operations in the LWR Fuel Cycle.* Washington, D.C. ERDA-76-43.

———. 1976b. *Draft Environmental Statement: Waste Management Operations, Savannah River Plant, Aiken, South Carolina.* ERDA-1537.

———. 1976c. *Draft Environmental Statement: Waste Management Operations, Idaho National Engineering Laboratory, Idaho.* ERDA-1536.

———. 1977a. *Alternatives for Long-Term Management of Defense High-Level Radioactive Waste, Savannah River Plant, Aiken, South Carolina.* ERDA-77-42.

———. 1977b. *Final Environmental Statement: Waste Management Operations, Hanford Reservation, Richland, Washington.* ERDA-1538.

———. 1977c. *Alternatives for Long-Term Management of Defense High-*

Level Radioactive Wastes, Hanford Reservation, Richland, Washington. ERDA-77-44.

General Accounting Office. 1977. *Cleaning Up the Remains of Nuclear Facilities—A Multibillion Dollar Problem.* Report to Congress. June 16. Washington, D.C. EMD-77-46.

Krugmann, Hartmut, and von Hippel, Frank. 1977. Radioactive Wastes: A Comparison of U.S. Military and Civilian Inventories. *Science* 197:883–85.

Lester, Richard K., and Rose, David J. 1977. The Nuclear Wastes at West Valley, New York. *Technology Review* 79:20–29.

Smith, Kirk R. 1978. Military Uses of Uranium: Keeping the U.S. Energy Accounts. *Science* 201:609–11.

JAN A. J. STOLWIJK

3 NUCLEAR WASTE MANAGEMENT AND RISKS TO HUMAN HEALTH

INTRODUCTION

In common with most technological undertakings the use of nuclear energy to generate electricity presents risks to human health. Some of these risks are associated with normal operations of reactors and the nuclear fuel cycle from mining through disposal of nuclear wastes, and others are associated with accidents, failures, breakdowns, and other abnormal and unplanned but conceivable events. There is a considerable amount of controversy surrounding the estimates of such risks, and there is perhaps even more disagreement concerning the acceptability of such risks. For a discussion of the difficulties in determining the acceptability of a risk to public health the reader is referred to Lowrance (1976). In considering the risk to human health of any major technological undertaking it is useful and even necessary to review the risks associated with not using this technology, or using alternative technologies to achieve the same ends. Such an analysis was recently carried out for the risks associated with nuclear power as compared with those of an equivalent amount of power generated by combustion of coal (NEPSG, 1977).

The determination of the potential risk to human health from nuclear waste materials is an exceedingly complex problem. The effect of nuclear radiation on human beings is dependent on the type and amount of ionizing radiation absorbed by the body and on the body area which absorbs it. High dose rates create a greater risk than is caused by the same total dose delivered over a long period of time. Nuclear waste material contains a large number of chemical species and radioisotopes with widely varying half-lives and radiation characteristics. As a result the age of the nuclear waste (since removal from the reactor) has a pronounced effect not only on the total radiation hazard but also on the likely type of interaction with humans, the mode of entry, and the target tissue. Ideally this chapter should provide an understanding of all these complex relationships.

In reality, within the current scope, it can only provide an overview of the most salient areas, such as the effect of ionizing radiation in biological tissues and the principles of dosimetry as they apply to humans. The most likely exposure to nuclear wastes is going to result in relatively low-level radiation exposure, and we will review the expected adverse health effects of such exposures. In addition we will consider the radiation characteristics of nuclear wastes at different times and possible avenues by which humans may be exposed to radiation from nuclear waste products.

EFFECT OF IONIZING RADIATION IN BIOLOGICAL TISSUE

A major characteristic of nuclear wastes is the intense radioactivity associated with it. When virgin nuclear fuel is first loaded into a nuclear reactor it consists mainly of natural uranium (^{238}U) somewhat enriched with ^{235}U, which is the fissile isotope. In course of the operation of the reactor the ^{235}U is gradually depleted, giving rise to fission products, the most abundant of which are isotopes of cesium (^{137}Cs) and strontium (^{90}Sr). In addition a number of heavy transuranic elements are formed of which ^{239}Pu is the one most discussed. The vast majority of these waste products are unstable, and after one or more radioactive disintegrations they become transformed into stable isotopes. Each radioactive isotope in nuclear waste material decays at its characteristic rate; a half-life can be less than a second or thousands of years. Each radioactive isotope also emits a characteristic radiation when it decays. This radiation is characterized by its nature: it can be electromagnetic, such as in X-ray or gamma radiation, or it can consist of particles, such as in alpha, beta, or neutron radiation. In addition each radioactive isotope emits at a characteristic energy level.

All radiation from radioactive isotopes is alike in its ionizing effect when the radiation interacts with any matter, including biological tissues. As a particle, such as an alpha particle or a beta particle, or very energetic electromagnetic radiation, such as gamma rays, pass through matter they can enter into collisions or in other ways cause such serious disturbances in the atoms that make up the matter traversed that large numbers of these atoms lose one or more electrons. This process is called ionization. Every such ionization uses up some of the original energy of the radioactive particle, until finally it has dissipated all its energy. An ionized atom can absorb a large amount of energy, and such an atom can easily break loose from the molecule of which it was a part. This can cause chemical changes in biological tissues. If such ionizations occur in metal or in a crystal with a relatively simple chemical and physical structure then the damage is not likely to be severe, but in biological tissue, with its chemical and structural complexity, such "hits" can have serious consequences. Since genetic information which governs function, structure, growth, and reproduction is contained in biological tissues it is clear that such ionization can have serious genetic consequences. It is likely that some of the damage in biological systems is repaired in time, especially if the dose rate and consequently the damage rate is low.

If a given ionizing particle interacts very strongly and thus produces many ionizations that are very close together, it is likely to produce more damage to biological tissues than the same total number of ionizations distributed over a larger tissue mass. As will be discussed later, a given dose received over a very long time or as a sum of partial doses received at intervals is thought to be less damaging than the same total dose received all at once in a short time interval.

The radiation dose is expressed in units of roentgen. This unit was originally defined as that amount of X rays or gamma rays that produces an amount of ionization totalling one electrostatic unit (e.s.u.) of either sign or 2×10^9 ion pairs in one cubic centimeter of air. This is also equivalent to 2.58×10^{-4}

coulomb per kilogram of air. For human dose measurements and estimates, the unit of dose is the rem. The rem (roentgen equivalent man) is based on the roentgen, but it incorporates appropriate adjustments for the relative biological effectiveness of different types of radiation and for a number of other complicating factors. For radiation doses in water and other materials the rad is often used as the unit of dosage. It is a unit of absorbed dose equal to .01 Joule/kg. (or 100 ergs/gram).

If a total amount of radioactivity is to be indicated, the unit of measurement is the curie, which was defined in Chapter 2 as that amount of radioactive isotope in which 3.7×10^{10} nuclear transformations occur each second. In terms of amounts considered acceptable to the human body as a whole body burden, the curie is a very large unit: the permissible body burden for tritium (^{3}H) is 10^{4} microcuries, and for plutonium (^{239}Pu) it is as low as 0.04 microcuries. The measure of radioactivity by curies is not directly useful for discussions of the health effects of radiation that originates outside the body, since the curie does not reflect the rate of absorption of radiation by the body. As an example, some smoke detectors contain 0.000000025 curies of radioactivity. If several people were moving about in a room with such a smoke detector, the actual exposure would depend on the time spent in the room, the distance from the smoke detector, and the screening by any objects between the source and the individuals.

Acute Effects and Long-Term or Delayed Effects of Ionizing Radiation

Exposure to a large dose of ionizing radiation produces adverse health effects of an acute nature; they manifest themselves in a period of from hours to weeks and lead to death or to recovery from the short-term illness. In addition to the acute effects suffered by survivors of a single, large-dose exposure, long-term effects can lead to genetic changes, which can be passed on to the next generation, or to somatic changes, which can cause neoplastic and other diseases years after the original exposure. When humans are exposed to much lower total doses over much longer periods, they also incur a risk of long-term genetic or somatic changes.

It is important to make a general distinction between the acute effects and the long-term effects. The nature of the risk is quite different for these two types of effects. When a large number of people are exposed to a short exposure of 1,000 rems of penetrating gamma radiation, all the individuals will suffer severe illness and almost all will die within a month. Each individual has a 100 percent probability of severe consequences, and the uncertainty is only in the degree of severity.

If a large number of people is exposed to an amount of radiation ten times that present as natural background for a period of thirty years, none will suffer acute effects of this exposure, and only very few will suffer genetic or somatic effects which are likely to cause illness and premature death. Most of the individuals so exposed would not suffer any observable effects, although the total dose would be in excess of 40 rems.

Ionizing Radiation in the Natural Background

Avoidable, man-made sources of ionizing radiation exposure should be considered in the perspective of the irreducible natural background exposure. This whole-body exposure continues throughout life and varies somewhat from location to location. All areas are exposed to cosmic radiation, which delivers a dose that depends on altitude and latitude: people in Florida receive a dose rate of 0.038 rem/year, and people in Wyoming receive 0.075 rem/year. The material of the earth's surface delivers a dose rate which varies depending on the composition of the material from as low as 0.015 rem/year in the southeast coastal plains to as high as 0.14 rem/year on the Colorado Plateau. Each human body contains naturally occurring radioisotopes which deliver an internal dose of 0.020 rem/year from potassium (^{40}K), 0.002 rem/year from carbon (^{14}C), and 0.002 rem/year from radon (^{222}Rn).

Man-made contributions to population exposures have been estimated in the BEIR report (NAS/NRC, 1972) as follows for 1970: medical exposure (mainly diagnostic X radiation), 0.072

rem/year; radiopharmaceuticals, 0.001 rem/year; radiation treatment of cancer, 0.005 rem/year; nuclear power, 0.000003 rem/year; global fallout from weapons testing, 0.004 rem/year (table 3.1).

It should be noted that these exposures are average exposures and certain areas and individuals will have exposures which are higher or lower than the average. Certain individuals receive much more diagnostic or therapeutic radiation, and people living near nuclear power plants receive more radiation than people living at great distances from such plants.

The UNSCEAR report of 1977 estimates world population exposures in rems per person per year as follows:

Natural sources	0.1
Commercial air travel	0.00011
Emission from coal-fired power plants	0.000005
Emission from nuclear power	0.00164
Fall-out from nuclear testing	0.00822
Medical diagnostics	0.0192

HEALTH EFFECTS OF ACUTE IONIZING RADIATION IN HIGH DOSES

In this section observable acute clinical effects of large doses of radiation are described. Persons subjected to these large doses are also subjected to delayed genetic and somatic effects which might not be observed until many years have elapsed.

Whole-body radiation doses below 25 rems do not produce observable clinical effects. Doses between 25 and 50 rems are likely to produce a decline in the number of white blood cells, but doses up to about 100 rems are unlikely to produce any additional immediate effects. At doses between 100 and 200 rems all irradiated persons will develop leukopenia (reduction in white blood cell count), and between 5 percent and 50 percent will experience vomiting about 3 hours after the exposure. No specific treatment is indicated and recovery will be complete after several weeks.

Whole-body exposure to between 200 and 600 rems will

Table 3.1 Population Exposure to
Ionizing Radiation in the U.S. in 1970

Source	Dosage (rems per year)			Total Dosage to U.S. Population (Person rems per year)
	Low	High	Average	
Cosmic radiation	0.038	0.075	0.044	9,000,000
Terrestrial radiation	0.015	0.150	0.040	8,200,000
Potassium-40	0.020	0.020	0.020	4,100,000
Carbon-14	0.002	0.002	0.002	400,000
Radon-222	0.002	0.002	0.002	400,000
Total background			0.108	22,100,000
Diagnostic X-rays			0.072	15,000,000
Radiopharma-ceuticals			0.001	200,000
Radiation therapy			0.005	1,000,000
Nuclear power			0.000003	600
Fallout			0.004	800,000
Total man-made			0.082	17,000,000
Total exposure			0.190	39,100,000

NOTE: Lifetime exposure = (0.190 rem/year) (70 years) = 13 rem. Reproductive lifetime exposure = (0.190 rem/year) (30 years) = 6 rem.

cause vomiting after about 2 hours. Victims will suffer severe blood changes, loss of hair after about 2 weeks, internal bleeding, and complicating infections. Therapy may consist of blood transfusions and antibiotics with a convalescent period for survivors of from 1 to 10 months. Depending on the treatment and other factors, mortality may range from 0 percent to 80 percent and will be caused by internal bleeding and massive infections.

Between 600 and 1,000 rems of whole-body exposure will cause vomiting within 1 hour; all other effects are the same as those caused by between 200 and 600 rems, except mortality will be between 80 percent and 100 percent and therapy will be much less effective. For doses above 1000 rems, there is no effective treatment. Vomiting is induced within 30 minutes, and death occurs between 1 hour and 14 days later, depending on the dose.

LONG-TERM AND DELAYED EFFECTS OF IONIZING RADIATION

Doses of ionizing radiation that produce acute radiation sickness also produce a risk in the exposed population that they or their offspring will experience delayed effects which might not manifest themselves for many years. Even acute doses too small to produce radiation illness and low doses delivered over very long periods by substantial increases in background radiation will produce an increased risk of delayed effects.

Our knowledge of these delayed genetic and somatic effects, especially at lower dose rates, is much less detailed and quantitative than would be required for firm assessments of these risks. Few human populations of large size have been exposed to a precisely known amount of radiation and have been followed over long periods of time to observe the long-term effects of the radiation. The long-term effect of ionizing radiation has been studied extensively in large animal populations under precisely controlled conditions, but extrapolating from results obtained in mice to predict the effects in humans is exceedingly difficult.

Long-term effects of radiation fall into three categories. The first of these is genetic effects, which occur in the form of gene mutations or in the form of chromosome aberrations. Both of these effects also occur spontaneously, and it should be recognized that small amounts of ionizing radiation do not cause new effects, but increase the rate at which already occurring effects take their toll. Almost every aspect of form and function of a living organism is determined to some extent by its genes, and it follows that a gene mutation can affect almost any aspect of an organism.

A second category is radiation effects that show themselves in the course of human development. A fetus *in utero* or a very young child exposed to substantial levels of ionizing radiation has a greater chance of suffering impaired growth or microcephaly or of being mentally retarded.

The third category of radiation effects is somatic effects, the most important of which is increased cancer induction. It is clear that acute effects really belong in the same category, but they have already been described and occur only at doses well above the range of concern here.

Genetic Effects of Ionizing Radiation

The most easily noted gene mutations are the single, dominant mutations associated with disease or abnormality, of which about one thousand have been described for humans. The spontaneous occurrence of a dominant mutation is seen in about one percent of births. These abnormalities include several types of anemia, dwarfism, extra fingers or toes, Huntington's chorea, and others. Recessive mutations are not as easily seen, since such a gene will only become visible if two recessive genes are inherited, one from each of the parents. There is a growing list of diseases caused by recessive mutations, which includes sickle cell anemia, cystic fibrosis, Tay-Sachs disease, and others. The spontaneous incidence of recessive diseases is usually estimated at 0.1 percent of all births.

Many dominant or recessive gene mutations govern changes in form or function which are acceptable and give little cause for concern. On the other hand many mutations have such severe effects that they are lethal and result in nonviable embryos. Such mutations cause apparent lowering of fertility, or increases in cases of spontaneous abortion, but they do not produce the same misery and suffering as do disease-bearing but viable mutations. Chromosome aberrations produce disease conditions in only 0.4 percent of all live births, but they are found in a very high proportion of spontaneous abortions.

Many cases of disease are not specifically the results of mutations but show evidence of complex hereditary causes. Surveys of the use of medical resources have shown that about 25 percent of all hospital treatment is for hereditary disease. It is not known whether complex hereditary diseases will increase in number or prevalence as a result of mutations induced by radiation. The incidence of inherited diseases with complex aetiology has been variously estimated at from 5.5 to 9.0 percent of the population.

A useful way to express the risk of genetic effects caused by ionizing radiation is to estimate the dose that will double the naturally occurring rate of spontaneous mutations. This estimate is referred to as the "doubling dose." This doubling dose is the hypothetical total dose that would be accumulated during a reproductive lifetime. Currently, about one percent of all persons born suffer from a serious genetic defect or disease. The BEIR report estimates that the doubling dose of chronic radiation is between 20 and 200 rems, and the UNSCEAR report estimates that the doubling dose is unlikely to be lower than 100 rems. These estimates imply that the natural radiation background which produces a dose of about 3–5 rems during the reproductive life of an individual contributes only a small fraction of the total number of spontaneous mutations. By the same reasoning, even smaller increments in ionizing radiation will produce even smaller numbers of mutations. The possible relationship between additional ionizing radiation and increases in general illness because of its mutational component was estimated by the BEIR report to be between 0.5 and 5.0 percent increase in general illness for an additional dose of 5 rems per generation. It should be noted that 5 rems per generation corresponds to a doubling of natural background radiation for the first generation so considered. Estimates of the dose to which the population will be exposed by all nuclear power operations run as high as 0.001 rem per year in the year 2000, or 0.03 rem for a reproductive lifetime.

Effect of Ionizing Radiation on Human Growth and Development

It has been known for many years that irradiation of a fetus *in utero* or of a very young child can have serious consequences in the child's subsequent development. Older data were not very systematic and consisted of observations of individuals who had received massive doses of 100 rems or more *in utero* when their mothers were treated with radiation for gynecological conditions.

More systematic knowledge about the dose rates required for subsequent reduced growth rates, mental retardation, and microcephaly was obtained from studies of the survivors in Hiroshima and Nagasaki and of the Marshall Islands. Individuals exposed *in utero* within 16 weeks after conception were more sensitive than individuals exposed later. Individuals exposed to about 25 rems showed some mental retardation. Those exposed to 50 rems showed retardation that was "profound": the individuals could not care for themselves, carry on a simple conversation, or perform simple calculations.

Irradiation after birth has an effect on development only at much higher doses of 700–1500 rems given over a period of time. Much less is known about the effect of low-level chronic irradiation on growth and development, but there are indications that chronic doses at 1 rem/day or 365 rems/year or less will not produce deleterious effects of this category. It is thus clear that other types of radiation effect will occur and require action long before one needs to be concerned about the effects on growth and development.

Somatic Effects of Ionizing Radiation

Excluding the acute somatic effects of large, single doses of ionizing radiation, which have been discussed earlier, the somatic effects manifest themselves years or decades after exposure. The effects are difficult to establish for two reasons: they are indistinguishable from similar diseases which occur spontaneously without irradiation and, since very few irradiated individuals develop diseases, a neoplasm in a given individual cannot be determined with certainty to be the result of ionizing radiation exposure. The effect of radiation, however, can be shown in the form of a greater number of cases in an exposed population than in an unirradiated population.

There are a number of somatic long-term effects, such as cataract formation and reduction of fertility, but by far the most important effect is cancer induction, especially if low-level chronic doses are to be the main concern. The most important form of cancer following irradiation is leukemia, but increased numbers of malignancies have also been found in the thyroid, lung, breast, bone, stomach, and intestinal tract.

Epidemiological data on which our knowledge of cancer induction by radiation rests were gathered in populations which were exposed to relatively high doses delivered in a relatively short time. One such population was British patients who received heavy doses of radiation for treatment of ankylosing spondylitis. Other populations were exposed in Hiroshima and Nagasaki. In most cases the dosage was of the order of 50 rems or higher, at a rate of 1 rem/min. or higher.

For these populations, increases in the incidence of all forms of cancer have been observed. The increased risk of cancer is in proportion with the excess in radiation exposure; although the precise relationship is not known, it is possible to develop estimates of the increase in cancer cases which will develop in a population exposed to ionizing radiation. These estimates are based on observations in populations who received doses of the order of 100 rems, and although some might argue that this would lead to exaggerated estimates at low doses and dose rates, the conservative hypothesis of zero threshold and linear dose-effect relationship should be adhered to. The estimates given in table 3.2 are taken from the 1977 UNSCEAR report.

In considering the risk estimates in table 3.2, it should be remembered that in the general population the "spontaneous" total cancer mortality is of the order of 200,000 per million population. The increase in cancer mortality due to 10 excess rems in a population of 1 million would therefore be 1000 more than the 200,000 cases to be expected as background cancer mortality.

It is clear from such considerations that epidemiological studies of large populations of 1 million or more who are thought to have received an excess dose of 1000 millirems are not likely to demonstrate an excess incidence of cancer associated with such an excess dose: the expected increase of 0.05 percent in total cancer mortality would not usually be statistically significant.

Table 3.2 Excess Cancer Mortality in a Population following Exposure to Ionizing Radiation in Excess of Natural Background. (Total Number of Excess Deaths per Million People Exposed per Rem)

Thyroid	10
Breast	50
Lung	25–50
Leukemia	20–50
Gastro-intestinal	10–15
All	100

Estimates or projections of the additional risks due to nuclear waste management are difficult and consequently must be regarded with considerable reservation because of the many unsubstantiated and even unstated assumptions which necessarily are made in the erection of such estimates. The U.S. Department of Energy in an Environmental Impact Statement (DOE, 1980) produced a number of such estimates based upon a once-through (no-reprocessing) fuel cycle and a fuel cycle which included reprocessing. These estimates were for the then-proposed program of geologic disposal, an alternative program in which ultimate disposal was delayed for thirty years, and for a no-action alternative. In addition, five assumed levels of nuclear power utilization were evaluated. All of these evaluations were made under the assumption of normal operations without any unplanned or accidental releases.

Such estimates should be regarded with limited confidence, but they do provide a relative perspective prepared on behalf of proponents of the maintenance and development of a nuclear option. Table 3.3 presents a summary of the estimates produced in the Environmental Impact Statement. If such estimates can be accepted as accurate within even a factor of one hundred, it is clear that the risks ought to be acceptable; the natural radiation background at individual residential locations in the U.S. varies by a factor greater than two without apparently causing concern to occupants.

Table 3.3 Estimated Fractional Increment to the Natural Background Lifetime Population Dose Commitment due to Normal Operation of Nuclear Waste Management Using Geologic Disposal

Nuclear Power Development Assumptions	Once-through Cycle		Reprocessing	
	Regional Dose Increment	Worldwide Dose Increment	Regional Dose Increment	Worldwide Dose Increment
Present capacity	2.0×10^{-5}	7.1×10^{-9}	1.0×10^{-3}	1.0×10^{-6}
250 gigawatt-electric by 2000	1.1×10^{-4}	3.5×10^{-8}	1.3×10^{-3}	1.7×10^{-6}
500 gigawatt-electric by 2040	1.9×10^{-4}	6.2×10^{-8}	4.6×10^{-3}	3.3×10^{-6}

HEALTH EFFECTS CONSIDERATIONS IN NUCLEAR WASTE MANAGEMENT

Without claiming to consider all possible eventualities we will consider in this section possible and plausible effects of normal, planned methods of waste management, processing, and permanent disposal as well as some unplanned but plausible mishaps, accidents, and catastrophic events.

Regardless of the methods of treatment and disposal the total amount of radioactivity in curies, the total weight of nuclear wastes, and the total amount of radiation emanating from these wastes is very large, and its very magnitude creates an enormous potential threat.

Possible exposures of large populations to ionizing radiation associated with nuclear wastes should be seen in relation to current regulatory controls and protection guidelines and natural background and medical exposure. There are a number of international and national commissions and agencies which set standards and rules for radiation protection and for the measurement of radiation levels. For the United States the most relevant organizations are the National Committee on Radiation Protection and Measurements (NCRP), the U.S. Environmental Protection Agency (EPA), and the Nuclear Regulatory Commission (NRC). For individuals who are occupationally exposed to ionizing radiation the maximum permissible dose is 5 rems/year of whole-body irradiation. If such a dose should be exceeded in any one year, then the permissible dose for that individual should be lowered in successive years.

The occupational dose to the skin should not exceed 15 rems in any one year; the hands should not receive more than 75 rems/year, and the forearms 30 rems/year. Pregnant women should be protected so that the fetus does not receive more than 0.5 rem. Life-saving actions by volunteers in an occupational setting should not put them at risk of whole body exposures of more than 100 rems.

Maximum permissible exposures for the general public are set at one tenth of the occupational limits; that is, 0.5 rem/year (in addition to natural background and medical exposures).

This is the exposure limit for any individual for whole-body irradiation. In addition to this individual limit there is an average population limit which is set at 0.17 rem/year from all sources other than natural radiation and medically administered radiation. The total dose from nuclear reactors is not permitted to be more than 0.005 rem/year. In addition to these general exposure limits there are very detailed sets of limits of concentrations of radionuclides in air and water and in the body, which are designed to limit internal exposures to safe levels.

All waste management procedures will be subject to all the applicable regulations cited above, and none will be licensed or undertaken until credible precautions have been reviewed and accepted.

A large expansion of the nuclear industry and its associated waste management, treatment, and disposal activities will still cause a risk to health in excess of current radiation risks, since a larger number of individuals will be subjected to a total dose somewhat greater than the natural background and the current medical dose, although as far as can be foreseen the permissible dose of 0.17 rem/year for the whole population will not be exceeded. The increased dose is, however, likely to be much lower than the current medical dose administered to the population. It is perhaps feasible to reduce the medical dose to the population by the use of improved technology and improved judgment criteria for diagnostic procedures. Such a reduction is likely to be greater than the increase in exposure caused by the orderly disposal of nuclear wastes.

In considering public health risks in normal and unplanned conditions surrounding nuclear waste management, it is useful to take into account some characteristics of the total system which are often not weighed.

Before ionizing radiation can have any effect on man either it must reach humans directly as radiation originating outside the body, or the radionuclides must be inhaled and absorbed in the respiratory system or absorbed with ingested food or water.

All our knowledge about the biological half-lives of various atomic species, the rate of concentration in various organ systems, and the radiosensitivity of different tissues is based on the maximum permissible concentrations of different radionuclides in the body and in air and water and food. The acceptability of these risks has been determined by groups of qualified and responsible experts who had no direct stake in nuclear development at a time when waste management had not yet become a controversial issue.

The permissible population dose for chronic exposure was lowered from the first recommendation but has since remained stable at current levels.

The measurement, monitoring, and verification of levels of ionizing radiation in the region of permissible doses is neither difficult, extraordinarily expensive, nor subject to disagreement. Therefore, excessive exposures of large or even small populations probably will not occur without complete awareness. If such exposures do occur they will be of quite limited duration before corrective measures can be applied.

External radiation exposures are normally avoided or controlled by interposing shielding materials between the radioactive material and people, or by increasing the distance between them, or by limiting the exposure time. Shielding from direct radiation for long-term storage and permanent disposal can be accomplished very effectively since 200 meters of overburden are totally effective in shielding the population at the surface. The remaining question of geological stability of the deep disposal area is not really relevant to the direct radiation problems: even if there is a major unexpected disruption, new shielding mass can be applied to prevent any radiation from reaching the surface. Direct radiation exposure from nuclear waste materials in operating reactors or during processing, temporary storage, or transport is prevented by containment, shielding, and aging of the fission products. Remaining direct radiation risks can be reduced by improving the integrity of containment structures, by siting away from population concentrations, and by transportation strategies which emphasize safety and the use of unpopulated routes. As an example, if transportation is by truck, knowledgeable safety personnel should accompany in a separate vehicle so that exposures following an accident can be kept to an irreducible minimum.

Venting of waste products into the atmosphere from permanent disposal sites does not appear to be very likely: even from very high yield underground test explosions, venting has been prevented effectively. Venting from processing plants and from operating reactors is a more plausible form of release of fission products to the atmosphere. In normal operations such emissions are kept well within permissible limits, and for accidental and catastrophic failures the release of gaseous fission products will be of minor concern compared with other exposures occurring at the same time.

Probably the most serious concern in the management of nuclear waste products is that of radionuclides being ingested with food and water and thus delivering an internal radiation dose. These could be dispersed from permanent disposal sites, from operating reactors, from temporary storage, or during transport. Ecological chains could concentrate radionuclides in food from very low concentrations in ground water and deliver radioactive dosages over wide areas and to large populations. Technical discussions of the relative protection afforded by deep burial in salt mines have appeared recently, and the reader is referred to these for a more detailed discussion of a very complex problem (Cohen, 1977; de Marsily et al., 1977).

In general it is well to remember that unavoidable risks can be reduced by appropriate planning: in most cases the risk of exposing the total population can be reduced substantially by selecting suitable locations for nuclear power plants, processing plants, and disposal sites and appropriate transportation times and routes. Estimates of risk, such as those in the Reactor Safety Report of the Nuclear Regulatory Commission or the Ford Foundation report "Nuclear Power, Issues and Choices," seldom take into account the effect of protective measures put into effect after an accidental release or of site planning that gives high priority to reducing population risks.

Should a large amount of nuclear waste material be dispersed on land following a catastrophic failure of containment or safety mechanisms, a sizeable land area may become unfit for human habitation, and agricultural production may be compromised over an even larger area. This could result in the abandonment of valuable property and much-needed agricultural productivity. There is likely to be pressure to resettle before this can be done safely and to raise the permissible level of radiation, since such a change would be immediately translated into widely distributed economic gains.

REFERENCES

Cohen, Bernard. 1977. The Disposal of Radioactive Wastes from Fission Reactors. *Scientific American* 236:21–31.

De Marsily, G.; Ledoux, E.; Barbreau, A.; and Margat, J. 1977. Nuclear Waste Disposal: Can the Geologist Guarantee Isolation? *Science* 197:519–27.

Department of Energy (DOE). 1980. *Management of Commercially Generated Wastes.* Washington, D.C. EIS–0046.

Lowrance, William W. 1976. *Of Acceptable Risk. Science and the Determination of Safety.* Los Altos, CA: William Kaufman, Inc.

National Academy of Sciences/National Research Council (NAS/NRC). 1972. *The Effects on Populations of Exposure to Low Levels of Ionizing Radiation.* Report of the Advisory Committee on the Biological Effects of Ionizing Radiations, Washington, D.C.

Nuclear Energy Policy Study Group (NEPSG). 1977. *Nuclear Power, Issues and Choices.* Cambridge, MA: Ballinger.

United Nations Scientific Committee on the Effects of Atomic Radiation (UNSCEAR). 1977. *Sources and Effects of Ionizing Radiation.* Report to the General Assembly, 1977. New York.

STANLEY M. NEALEY AND JOHN A. HEBERT

4 PUBLIC ATTITUDES TOWARD RADIOACTIVE WASTES

Public attitudes are certain to influence the future of nuclear power in the United States; how they will influence decisions about radioactive waste management is not yet clear.[1] Though a substantial segment of the public is opposed to the continued development of nuclear power, nuclear waste disposal *per se* is not in dispute. Many people disagree, however, about management alternatives and siting decisions. The purpose of this chapter is to review what is known about public attitudes and beliefs about nuclear power and radioactive wastes.

The analysis is organized as follows: in the first section the emergence since 1960 of radioactive wastes as a public opinion issue is examined and the areas in which public concern has been focused are explained. The second section is a discussion of how the public weighs different priorities in waste disposal, such as cost versus safety, and shows how environmentalists differ from nuclear engineers in these priorities. Other major differences in attitudes toward radioactive waste among various groups in the society are detailed in the third section. The fourth section is an analysis of the extent of public knowledge about radioactive wastes and their management and of the role of the mass media in providing such information. Several implications for public policy that follow from public attitudes about radioactive wastes are discussed in the concluding section.

1. See Chapter 6 for a fuller discussion of this issue.

EMERGENCE OF RADIOACTIVE WASTES AS A PUBLIC OPINION ISSUE

In 1960, a nuclear plant siting survey at Indian Point, New York, conducted by the Opinion Research Corporation (1960) asked residents if they felt confident about the safety rules set by the Atomic Energy Commission for the disposal of atomic waste. Fifty-seven percent felt confident, 13 percent had some question, and 30 percent said that they didn't know. In several other studies between 1960 and 1973, not a single respondent spontaneously brought up the subject of radioactive wastes when asked for reasons why he or she opposed nuclear power. It seems likely, therefore, that the issue of radioactive wastes was not very salient to the public prior to about 1973.[2]

In 1973, however, a leak of 115,000 gallons of high-level liquid radioactive waste from the nuclear weapons facility at the Hanford Reservation in Washington State received wide publicity. National Harris polls conducted in 1975 and 1976 found that 16 and 14 percent of people opposed to nuclear power development mentioned problems with radioactive waste disposal as a disadvantage of nuclear power. Reactor leaks and accidents were other specific dangers mentioned by about 15 percent of nuclear power opponents, but most common was the unspecific perception that nuclear power was "dangerously unsafe or a health hazard" (Rankin and Nealey, 1978a).

A somewhat different picture of the importance of radioactive wastes emerged, however, when people surveyed were given a list of purported problems or disadvantages to pick from. Radioactive wastes then jump to the number one position among nuclear concerns. As early as 1974, 52 percent of the people in one survey (Opinion Research, 1974) checked radioactive waste as a "serious problem" of nuclear power. "Radiation discharge" and "nuclear accidents" were also on the list,

2. Although nuclear attitude surveys have been conducted for over thirty years (see Kay and Gitlin, 1949), it was not until 1960 that questions about radioactive waste disposal were included.

but they were checked by less than half as many people. In Harris studies (Harris, 1975, 1976) 63 and 67 percent of the people surveyed checked radioactive wastes as a "major problem" associated with nuclear power. Explosions, accidents, and possible meltdowns were less frequently chosen.

Concerns about radioactive wastes by no means translate directly into opposition to nuclear power. In national surveys taken by Cambridge Reports (Cambridge Reports, 1975, 1976) only 22 and 26 percent of the general public endorsed the statement "The radioactive wastes from nuclear power plants are so dangerous that we cannot afford to risk producing them, since we cannot insure that in 100 or 1,000 years from now there won't be a leak." Approximately half believed that modern technology could find a safe way "to store nuclear wastes so there will never be a problem." In 1978, Harris found that 28 percent of respondents agreed with the statement that we should "forget about nuclear power today because we cannot predict with absolute certainty what might happen to this spent fuel and wastes in many hundreds, even thousands, of years."

Much of the concern about radioactive wastes obviously involves citizens' faith in the scientific-engineering adequacy of waste disposal technologies. Residents in the state of Washington were asked in 1977 for their reaction to the statement "Scientists already know how to dispose of radioactive wastes, so let's get on with it." Twenty-two percent agreed with this statement; 41 percent disagreed (Rankin and Nealey, 1978b). A 1978 national Harris poll found that 42 percent of the public did not believe that the U.S. had the know-how to store radioactive wastes until they decay to harmless levels.

The public is generally agreed on two further issues: the need for ongoing technical monitoring of radioactive waste repositories and the need to isolate wastes from human populations. In a 1978 study (Lindell et al.), a substantial majority of those asked agreed that frequent monitoring of disposed wastes "will be essential to assure people's health and safety." There was substantial disagreement, however, as to whether such monitoring would, in fact, be able to insure long-term safety. In the same study, 32 percent said that they would be unwilling to live within 100 miles of a waste repository, yet 29 percent expressed willingness to live within 10 miles of such a facility.

Perhaps the most significant point about public opinion on radioactive wastes is the widespread lack of familiarity with the issue. Even in the late 1970s as many as 36 percent of those interviewed declined to express an opinion on questions about radioactive wastes. In contrast, only about 17 percent fell into the "don't know" category in mid-1970s surveys asking about overall support for or opposition to nuclear power, and even fewer lacked an opinion on that subject following the Three Mile Island accident. Regarding radioactive waste, even among those with an opinion, knowledge about the subject is quite limited.

In sum, since the 1973 Hanford leak that brought radioactive wastes to widespread public attention, waste disposal has been the issue most frequently picked as the problem with nuclear power. A substantial part of the public doubts that adequate waste disposal techniques are now known, and a growing number mention radioactive waste problems as a reason for halting further nuclear development. In the absence of prompting, however, radioactive waste appears not to be the first concern that comes to mind in connection with nuclear power, perhaps because reactor safety issues can evoke more vivid visions of catastrophic accidents. Finally, radioactive waste remains a relatively obscure topic for a significant segment of the public.

PUBLIC PRIORITIES IN RADIOACTIVE WASTE DISPOSAL

How much is the public willing to pay for safe disposal of radioactive wastes? What level of safety is desired? Is short-term safety or long-term safety more important? Because trade-offs among priorities are inevitable in the course of making decisions about radioactive waste, it is important to know what weight the public gives to these and other values.

To help overcome respondents' lack of familiarity with radioactive waste issues, one study (Maynard et al, 1976) presented a 22-minute film describing, in a neutral way, basic information about radiation and the technical options for radioactive waste disposal. The respondents' task—to make priority judgments about what a good waste management system should emphasize—was explained in the film. Respondents were asked to judge the relative importance of four factors: cost (passed on to the electricity consumer), short-term safety (risks to current generations), long-term safety (risks to future generations), and accident detection and recovery (the ability to monitor and respond to problems with the disposal system if they should occur).

Reactions to these issues are shown in figure 4.1. Members of environmental groups ranked long-term safety as most important, with about 60 percent ranking it number one. Short-term safety was next in importance for environmentalists, and cost was given little importance. The environmentalists seemed to be saying, "It must be safe in the long term whatever the cost."

The answers given by nuclear engineers were quite different. As shown in figure 4.1, short-term safety was their prime

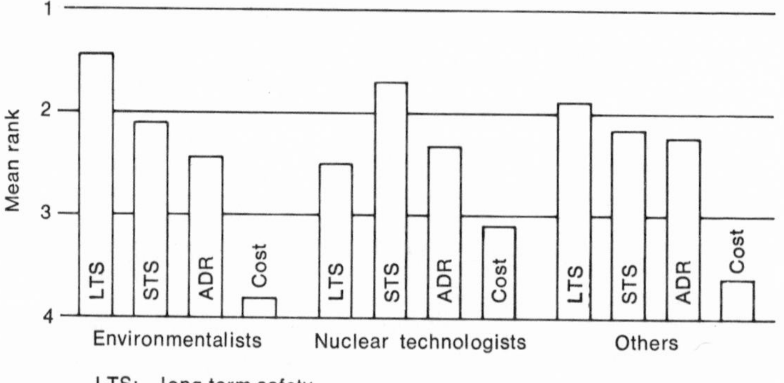

Figure 4.1. Mean ranks on the four waste disposal factors (adapted from Maynard et al., 1976)

concern, with long-term safety third. While cost was also last in importance for nuclear engineers, some 29 percent ranked cost either first or second. Only one of the environmentalists ranked cost first, and three percent ranked it second.

The others surveyed (members of civic organizations, business groups, and women's organizations and students) were probably more representative of the general public. They tended to give long-term safety most emphasis. Cost was again last, but not so far down in importance as it was for environmentalists.

In the same study, respondents were asked to judge "How safe is safe enough?" The average respondent rated ten deaths per year from U.S. radioactive wastes in the short-term (the present generation that uses nuclear power) as a level of safety that would fall in the middle of a ten-point scale ranging from "satisfactory" to "very unsatisfactory." Over the long term (future generations), one death per year was given a middle rating, indicating a more stringent requirement for long-term safety. A $5 per month charge per family for waste disposal was rated in the middle of the satisfaction scale, indicating high willingness to pay.[3] An accident detection and recovery system that was effective enough to reduce deaths by 75 percent in the event that something went wrong was rated in the middle of the satisfaction scale.

Satisfaction with various levels of long-term safety are shown in figure 4.2. The nuclear technologists were more easily satisfied than the "other" groups, which are probably more typical of the American public than are nuclear engineers. Environmentalists, on the other hand, appear much harder to satisfy. The nuclear technologists would be about as satisfied with one death per year in the long term as would environmentalists with one death in a hundred years. Data like these make it easier to understand why these two groups are often at loggerheads. It is also interesting to note that both of these groups were about equally far out of the mainstream (in different directions) as represented by the other respondents.

3. Data collected in 1975.

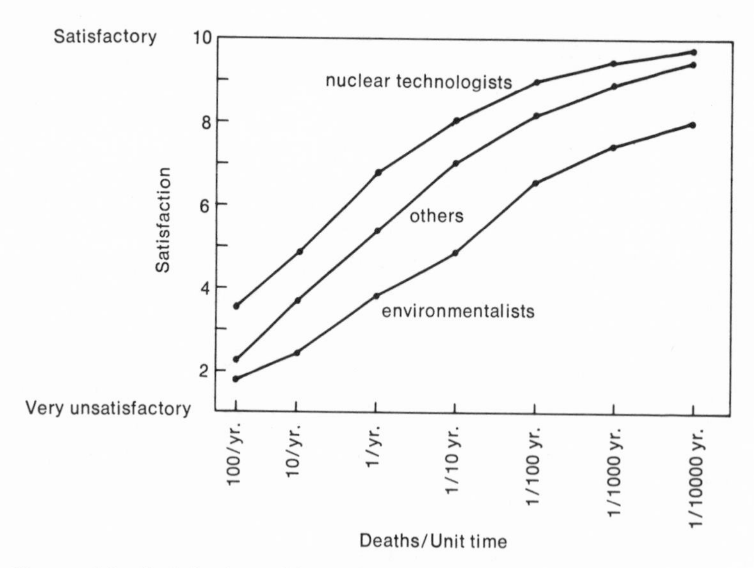

Figure 4.2. Satisfaction with performance of waste disposal for long-term safety

Part of the difference between environmentalists and nuclear technologists appears in their perceptions of the broader energy situation. Nuclear engineers reported that they did not believe that conservation and other alternatives to nuclear energy would be able to meet essential U.S. energy needs. Environmentalists tend to believe the opposite, so it is not surprising that the two groups evaluate the costs and risks of radioactive wastes differently. If nuclear power is essential for meeting fundamental needs, then the risks in radioactive waste management become more worth bearing.

What about the differences in perceptions concerning radioactive waste management in general? The nuclear technologists saw other social problems as more important than radioactive wastes in both the present and the future. This is no doubt based on their assessment that radioactive wastes are a problem that can be solved, even though it requires very careful handling. Since nuclear engineers are personally in positions to help solve the problem, they may feel more in control of it, much as many people feel more secure when driving a car rather than when being driven.

GROUP DIFFERENCES IN ATTITUDES

Most surveys show from 10 to 20 percent more men than women in favor of nuclear power (Melber et al., 1977). Middle-aged and older people tend to favor nuclear power somewhat more than the young. Support for nuclear power increases with years of education and with income level. There is also a slight tendency for nuclear power to be favored more in the West than in the Northeast. Given the tendency for attitudes toward radioactive waste to resemble attitudes toward nuclear power in general, it is not surprising that similar differences among demographic groups are found in studies of attitudes toward radioactive waste.

In three 1975 and 1976 Cambridge Reports national surveys, an average of 59 percent of men believed that technology could solve the radioactive waste problem, while only 42 percent of women had such faith. Respondents with more education and higher income were more likely to believe that technology would find a way to safely dispose of radioactive wastes. About 55 percent of middle-aged respondents expressed faith in technology, while only 48 percent of those older and younger held this belief. As is true for attitudes toward nuclear power, women, the less educated, the less affluent, and the younger respondents tended more frequently to select the "not sure" answer when asked about their prediction whether technology could solve the waste problem.

People who live near nuclear power plants tend to have more positive attitudes toward nuclear power in general, and radioactive wastes in particular, than people who are not "nuclear neighbors." In the 1975–1976 national surveys mentioned above, 53 percent of those living near existing or planned nuclear plants expressed faith that technology would solve waste problems, while only 46 percent in areas without nuclear plants held such an optimistic view. In a survey in the state of Washington (Rankin and Nealey, 1978b), residents near the Hanford nuclear installation were less convinced than the average state resident that radioactive wastes were "one of the most serious threats facing the world" or that radioactive wastes were a "very important" problem compared with others that future generations may face. "Nuclear neighbors" tended to agree more than the average respondent that "the benefits of nuclear power are much greater than the risks of nuclear waste storage and disposal." These attitudes are especially striking considering the great quantities of waste stored at the Hanford Reservation, some of which has leaked and has received extensive attention by the news media.

As might be expected, environmentalists in this Washington State study showed considerable concern over radioactive wastes. This group disagreed strongly that the benefits of nuclear power are "much greater" than the risk from radioactive wastes. They tended to agree that nuclear power plant construction should be stopped pending solution of the waste problem.

At least part of the disagreement between those who favor nuclear power and those who do not was found to be rooted in the different weights they attach to basic life values. Pronuclear respondents place more importance than neutral or antinuclear respondents on "a comfortable life," "family security," and "national security." In contrast, antinuclear respondents give greater emphasis to "a world of beauty" and to "equality." The general sample of Washington State residents and the nuclear neighbors placed family security ahead of all other values. Environmentalists ranked family security in fifth place and put "a world at peace" at the top of their list of important values. Generally, antinuclear respondents saw nuclear power as thwarting highly-held values, while pronuclear respondents saw nuclear power as promoting their highly-held values.

PUBLIC KNOWLEDGE ABOUT RADIOACTIVE WASTE

How great is the risk from radioactive wastes? Can these risks be managed safely? How can this best be done? Public knowledge about these and other radioactive waste questions is not well developed, partly because of the newness of waste management as a topic of public discussion. This section explores both what the public knows and how it gets its information, including the role of the mass media.

In the Washington State study cited above, 13 multiple choice questions were used to measure knowledge about radioactive wastes and their disposal. The average respondent got only 3.64 answers correct, just a little better than the performance that could have been obtained from random guessing. Pronuclear respondents averaged 3.84 correct compared with 3.38 for antinuclear respondents. Those living near the Hanford Reservation in eastern Washington and members of environmental interest groups were more knowledgeable than the general public. Nuclear neighbors scored 4.79 answers correct, and the environmentalists averaged 4.68 correct.

Asked how long it would take for 99 percent of the long-lived radioactivity to disappear from radioactive wastes prepared for permanent disposal, for example, 22 percent checked "10,000 years" and 30 percent checked "100,000 years or longer." The most nearly correct figure is 1,000 years. Nineteen percent got the question right, while 29 percent underestimated the time involved. On other questions, 73 percent overestimated the volume of radioactive wastes that would represent one person's lifetime share if all electricity were produced by nuclear power. About a third of the respondents did not understand that reprocessing waste is necessary to produce an atomic bomb; about half of those opposed to nuclear power held this misconception. A majority mistakenly believed that commercial wastes from power plants are currently in liquid form, and a plurality thought that liquid wastes could not be solidified. While 44 percent knew that long-term disposal plans call for deep geologic isolation, 35 percent believed that storage in tanks just below the earth's surface would be a permanent method.

Since respondents generally did not distinguish between military and commercial wastes, it seems likely that the high attention given by the media to the way military waste is stored on the Hanford Reservation and the associated leaks are responsible for some of the misconceptions found in the Washington State survey (Rankin, Nealey, and Montano, 1978). While most of the misinformation would lead to overestimating the hazards of radioactive wastes, there was one exception: respondents underestimated the severity of the 115,000-gallon waste leak at Hanford in 1973 and thought that it was discovered more promptly than it actually was.

Where does the public get what little information it has about radioactive wastes? The mass media are the obvious sources. Washington State residents listed newspapers, television, and news magazines, in that order, as their prime sources of nuclear information. Environmentalists, however, considered environmental groups their most prominent source, and nuclear neighbors relied on government agencies second behind newspapers. "Neutral" respondents, found to score significantly lower on the nuclear knowledge test than either pronuclear or antinuclear respondents, relied most on television.

A somewhat different picture emerges when the public is asked which source of nuclear news they trust most. Harris polls in 1975 and 1976 found "network TV news" in the top spot, trusted most by 25 percent of respondents nationwide. Next came "news magazines," with 22 percent, and "daily newspapers," with 16 percent. In the state of Washington, however, "government agencies" were trusted more than the "news media," with "utility companies" and "environmental groups" in third and fourth places as sources of trustworthy information about nuclear power. However, none of these sources scored above the middle of a "trust" scale. This skepticism is similar to that voiced by respondents to a Becker Research Corporation national survey (Becker, 1973, 1974): by a margin of 62 percent to 17 percent, people thought that the news media tend to emphasize the problems but not the reliable performance of nuclear power plants.

How do the media actually perform? Rankin and Nealey conducted a study of 266 stories related to nuclear power aired by the three major television network news shows from June, 1972, through December, 1977 (Rankin and Nealey, 1979). The average air time was only 77.3 seconds. A hypothetical viewer equipped with three television sets who watched all the network news every night would have received only 62 minutes of information per year about nuclear power. Radioactive wastes were a major topic in only 17 of the 266 television shows; it was a minor topic in an additional 15. While news is only one aspect of television programming, these data indicate that television did little to inform the public about radioactive wastes, at least up to the late 1970s.

Nor have print media given extensive attention to radioactive wastes. In one study (Rankin, Nealey, and Montano, 1978) of nuclear stories in magazines and newspapers, radioactive wastes were the main topic of less than five percent of the stories sampled.

Why is so little attention given to radioactive wastes, a subject that respondents in nuclear surveys rank ahead of any other nuclear topic as an area they want to learn more about? The media tend to cover events, not issues. But except for relatively uncommon waste leaks, radioactive waste is an "issue rich-event poor" topic. Although reactor safety stories have increased in the wake of Three Mile Island, radioactive waste information is still relatively sparse because there are so few events around which the media can construct radioactive waste stories.

Does the media tend to present a relatively unbiased view of nuclear power and radioactive wastes, or is the news slanted? Several studies (Rankin, Nealey, and Montano, 1978) of printed and televised news stories about nuclear power have found a tendency for the media to discuss costs and risks a good deal more than the benefits of nuclear power. About twice as many stories would tend to push a neutral reader or television viewer in an antinuclear direction than would tend to push such a person in a pronuclear direction.

Regarding radioactive wastes more specifically, of 13 newspaper articles sampled in the same study, one was rated pronuclear, three neutral, and nine (69 percent) antinuclear. Five discussed only costs, and four discussed mostly costs but some benefits. None discussed costs and benefits equally.

Both print media and television news have presented nuclear power in a more negative light than coal (Rankin, Nealey, and Montano, 1978), in spite of the severe health problems associated with coal combustion. In contrast, television news has presented solar power in very positive terms (Rankin and Nealey, 1979). Of course, these patterns do not automatically indicate media bias, for there is no rule of journalism that says the good news and bad news about every topic must be balanced. Nor does the tendency of the media to emphasize the controversial nature of nuclear power constitute evidence that the mass media created the controversy over nuclear power. Still, the pattern deserves to be noted.

In summary, the public is poorly informed about radioactive waste issues. It seems safe to say that the Hanford waste leak of 1973 is the only radioactive waste event of enough salience to be recalled by any significant number of the public. Considering how little attention has been devoted by the media to nuclear wastes, it is perhaps not surprising that the public's information about radioactive wastes and their management is not very good.

CONCLUSION

Three interesting questions emerge from the material reviewed in the chapter: (1) How much of the concern about radioactive wastes is genuine, and how much is part of a strategy to derail further development of nuclear power? (2) How might the public become better informed about radioactive waste issues? and (3) To what extent would improved knowledge lead to changes in attitudes about overall nuclear development? This section attempts to answer these questions and suggests a broader implication for policy making that emerges from what we know about public opinion on radioactive wastes.

Genuine Concern or Antinuclear Strategy?

A number of writers have argued that basic differences in orientation toward further economic growth and toward life styles are at the root of opposition to nuclear power (e.g., McCracken, 1977; Tucker, 1979). This thesis holds that opposition to nuclear power is a strategic ploy in a larger political and social struggle to change the U.S. way of life, to emphasize conservation and smallness rather than energy production, growth, and bigness. Amory Lovins's argument (Lovins, 1977) for an energy future based on solar power and intense conservation is often mentioned as the rallying point for this purported social movement.

Because radioactive waste disposal is a salient and essential feature of nuclear power, failure to develop a solution acceptable to the American public will eventually bring nuclear development to a halt. So opposition to nuclear waste activities and plans could be strategic in the sense that it might support a roadblock to nuclear power. If so, resolution of the waste issue without acrimony would not be likely, since nuclear opponents would see it as giving the green light to further nuclear development.

We know of no attitude data that bear on this speculation other than the indirect evidence presented above on the differing values of those who support and oppose nuclear power. For some individuals, radioactive wastes have probably functioned as a "justification issue." That is, some people who oppose nuclear power are not able to articulate just why, so when radioactive wastes are listed in an opinion survey as a possible disadvantage of nuclear power, it is as though some respondents say, "Ah, yes! That's why I oppose nuclear power!" Still, it seems very unlikely that the concerns about nuclear power expressed by such a large percentage of citizens from many different social backgrounds could be much influenced by any sort of deliberate strategy. Therefore, there is every reason to think that, for many, the concerns are real. Further, as we have indicated, a majority of the public appears to believe that a technical solution is both possible and probable, though not yet at hand.

Reducing Public Ignorance

Few members of the general public are very knowledgeable about radioactive waste issues, and misconceptions are prevalent. The Department of Energy has a responsibility to provide information to the public about all energy technologies. Although a sizeable portion of the public considers government agencies a credible source of information, the public, except for the neighbors of the Hanford Reservation, reports having received little information about nuclear technology from the government. Perhaps as a consequence, the public does not look to government agencies as a good source of information. Moreover, the government's energy brochures to date have been boring and difficult to read. Taken together, this suggests that the government's information programs are ineffective.

But some improvement in public knowledge may be possible; we have cited evidence that the public wants to become better informed about radioactive wastes. Since the mass media are widely perceived as the best way to get information, it seems obvious that the Department of Energy might make more intense efforts to provide media sources with information about radioactive waste technology, government plans, and waste-related issues. DOE might go even further and create waste disposal "events" suitable for news coverage, in recognition of the fact that the media tends to give space to events rather than to issues. While the gains in public knowledge are certain to be modest, such steps would at least encourage a more active media role in informing the public about radioactive wastes.

Would Improved Knowledge Change Attitudes?

Four considerations bear on this question. First, the link between knowledge and attitude is weak on most public issues. Regarding radioactive wastes, while pronuclear respondents in one study were slightly more knowledgeable than antinuclear respondents, no significant difference in knowledge was found between pronuclear "neighbors" and antinuclear environmentalists. It is obvious that spokespersons for antinuclear groups manage to obtain good command of the facts about nuclear power, yet they are not "converted."

Second, the link between attitudes about radioactive wastes and attitudes toward nuclear power is rather weak. Even supporters of nuclear power express concern about waste disposal problems, but this concern does not lead them to oppose further nuclear development. Analysis of the 1975 and 1976 Harris surveys by Melber et al. (1977) suggests that attitudes toward radioactive wastes are less accurate at predicting overall nuclear support or opposition than are five other types of nuclear attitudes, the most important of which is toward power plant safety.

Third, those who oppose nuclear power do so for a variety of reasons, of which waste disposal is only one. Concern over a number of nuclear issues is closely interrelated for many people, so the unique impact of any one issue is difficult to measure. But power plant safety long has been the first worry spontaneously mentioned to opinion pollers, and the Three Mile Island accident probably has brought plant safety into even greater prominence. Resolution of waste disposal issues won't obviously lead to lowered concern about plant safety, and therefore may not change nuclear power attitudes.

Finally, radioactive wastes have the distinction of being the one nuclear issue on which the adversaries substantially agree. Nuclear supporters have been only a little less concerned with waste disposal problems than have nuclear opponents. Given this unusual status, radioactive wastes conceivably could prove a pivotal battle in the war over nuclear development. For the present, a majority of the public expresses faith in the ability of nuclear technologists to develop reliable waste management technologies. Should this faith be shaken profoundly, as by a waste disposal accident analogous to the Three Mile Island plant safety incident or by lengthy delay in resolving waste disposal problems, then some of the current support for nuclear development would be seriously undermined.

In sum, there is not much reason to believe that improved public knowledge about radioactive waste management would lead to a significant increase in overall public approval of further nuclear power development. But if nuclear supporters perceive that their concerns about waste disposal cannot be taken care of, then a decrease in nuclear approval would likely result.

Implications

An important implication follows from the information discussed above and earlier in this chapter. Nothing approaching a consensus in public opinion is likely on the upcoming policy decisions about radioactive waste management. Among numerous factors conspiring against the development of such agreement, four deserve particular mention.

First, uncertainties about long-term safety and accident detection and recovery cannot be resolved completely prior to actual experience, so substantial disagreement is likely to continue among both experts and the informed public; this will trickle down to the remainder of the public. Second, segments of the public have very different basic life values that lead to divergent perceptions of the broad energy situation; this leads to irreconcilable conflicts over the risks associated with radioactive waste management. Third, even if government begins to work with the mass media more skillfully, the public is likely to be exposed to a somewhat unbalanced diet of media stories that reinforce skepticism about nuclear power generally and radioactive waste management in particular. Finally, even if greater media attention is devoted to radioactive wastes, the issue will probably continue to suffer the fate common to most nonpocketbook issues—low public awareness with relatively few people having much factual knowledge of the issue.

Because of both the nature of the radioactive waste problem and the nature of public opinion on it, then, there is little likelihood that government policy makers can bring much consensus behind any conceivable waste-management plan. Hence, decision making on radioactive waste probably should not be delayed in the false expectation that such a consensus can be built.

REFERENCES

Becker Research Corporation. 1973, 1974. In Melber, B. D.; Nealey, S. M.; Hammersla, J.; and Rankin, W. L. *Nuclear Power and the Public: Analysis of Collected Survey Research.* PNL-2430. Seattle: Battelle Human Affairs Research Centers, 1977.

Cambridge Reports, Inc. 1975, 1976. In Melber et al., 1977.

Louis Harris and Associates, Inc. 1975. *A Survey of Public and Leadership Attitudes toward Nuclear Power Development in the United States.* Report conducted for Ebasco Services, Inc. In Melber et al., 1977.

______. 1976. *A Second Survey of Public and Leadership Attitudes toward Nuclear Power Development in the United States.* Report conducted for Ebasco Services, Inc. In Melber et al., 1977.

______. 1978. In Melber et al., 1979.

Lindell, M. K.; Earle, T. C.; Hebert, J. A.; and Perry, R. W. *Radioactive Wastes: Public Attitudes toward Disposal Facilities.* B-HARC-411-004. Seattle: Battelle Human Affairs Research Centers.

Lovins, Amory B. 1977. Energy Strategy: The Road Not Taken? *Foreign Affairs* (1977): 65–96.

McCracken, Samuel. 1977. The War Against the Atom. *Commentary* (September 1977): 33–47.

Maynard, W. S.; Nealey, S. M.; Hebert, J. A.; and Lindell, M. K. 1976. *Public Values Associated with Nuclear Waste Disposal.* BNWL-1997. Seattle: Battelle Human Affairs Research Centers.

Melber, B. D.; Nealey, S. M.; Hammersla, J.; and Rankin, W. L. 1977. *Nuclear Power and the Public: Analysis of Collected Survey Research.* PNL-2430. Seattle: Battelle Human Affairs Research Centers.

Melber, B. D.; Nealey, S. M.; Weiss, C. S.; and Rankin, W. L. 1979. *Nuclear Power and the Public: Update of Collected Survey Research.* B-HARC-411-020. Seattle: Battelle Human Affairs Research Centers.

Opinion Research Corporation. 1960. In Melber et al., 1977.

______. 1974. In Melber et al., 1979.

Rankin, William L., and Nealey, Stanley M. 1978a. Attitudes of the Public about Nuclear Wastes. *Nuclear News* 21:112–17.

______. 1978b. *The Relationship of Human Values and Energy Beliefs to Nuclear Power Attitudes.* B-HARC-411-007. Seattle: Battelle Human Affairs Research Centers.

______. 1979. *A Comparative Analysis of Network Television News Coverage of Nuclear Power, Coal, and Solar Stories.* B-HARC-411-005. Seattle: Battelle Human Affairs Research Centers.

Rankin, W. L.; Nealey, S. M.; and Montano, D. E. 1978. *Analysis of Print Media Coverage of Nuclear Power Issues.* B-HARC-411-001. Seattle: Battelle Human Affairs Research Centers.

Tucker, William. 1979. Environmentalism: What Is It Really About? Talk presented at the Atomic Industrial Forum meeting, Kansas City, MO, February 1979.

PAUL SLOVIC AND BARUCH FISCHHOFF

5 HOW SAFE IS SAFE ENOUGH? DETERMINANTS OF PERCEIVED AND ACCEPTABLE RISK

Citizens of modern industrial societies are learning a harsh and discomforting lesson: the benefits from technology must be paid for not only with money, but with environmental degradation, anxiety, illness, injury, and premature loss of life. Through the news media, the American public has experienced a relentless parade of new and exotic hazards. As Rabinowitch (1972) observed:

> One day we hear about the danger of mercury, and run to throw out cans of tuna fish from our shelves; the next day the food to shun may be butter, which our grandparents considered the acme of wholesomeness; then we have to scrub the lead paint from our walls. Today, the danger lurks in the phosphates in our favorite detergent; tomorrow the finger points to insecticides, which were hailed a few years ago as saviors of millions from hunger and disease. The threats of death, insanity and—somehow even more fearsome—cancer lurk in all we eat or touch. [p. 5]

The daily discovery of new threats and their widespread publicity threatens to create a national neurosis characterized by mistrust of technology and obsessive preoccupation with risk. It is in this context that the debate over nuclear waste is set.

We are indebted to Sarah Lichtenstein, Leroy Gould, and Scott Slovic for their helpful comments on earlier drafts of this chapter. This work was supported by the National Science Foundation under Grants ENV77–15332 and OSS–16564 to Perceptronics, Inc. Any opinions, findings and conclusions, or recommendations expressed in this publication are those of the authors and do not necessarily reflect the views of the National Science Foundation.

One reaction to these perceived hazards is expressed by writer Elizabeth Gray (1976):

> I have dealt with the situation basically by trying to drop out. I go to my doctor as seldom as possible; I take as few medications as possible. I . . . go into my supermarket, buy non-additive bread, take my vegetables and fresh fruits home and scour them to attempt to get the pesticides off.
>
> Now when you come at me with a nuclear decision, I think a person like me is . . . going to be propelled into the picket lines when you want to invade the very precarious private space that I am trying to forge . . . to protect myself from . . . [the] very dangerous effects of my society.
>
> And when you begin to invade my space with nuclear plants, people like me will say, "No way. Someplace else. . . ." [p. 200]

The option of dropping out gives us some control over the level of technological risk to which we are exposed. However, reduction of risk typically entails reduction of benefit. Where individual control is possible, each of us must cope with such dilemmas by personally weighing the costs against the benefits. Other dilemmas, such as the management of nuclear wastes, can only be resolved by society as a whole.

The urgent need to help society cope with risks has produced a new intellectual discipline, "risk assessment" (Kates, 1978; Lowrance, 1976; Otway and Pahner, 1976; Rowe, 1977; Schwing and Albers, 1980). Risk assessment aims to determine the seriousness of a hazard and whether society should be exposed to it. Determining this requires an extraordinary degree of cooperation between the sponsors of technology, the public, the representatives of the public, and specialists from many fields. Technical issues require the efforts of physicists, biologists, chemists, and engineers. Social issues involve lawyers, political scientists, geographers, sociologists, economists, and psychologists. Specialists in decision making attempt to coordinate this diverse expertise. They ask, in effect, this question: "Given our society's values and all this knowledge, what actions should be recommended?"

However valid the assessments of specialists might be, the decisions eventually made by society will reflect social and political pressures as well as the calm, analytic weighing of costs and benefits. Before acting, participants in those decisions must engage in an intellectual process to which risk assessment may make an important contribution. They must judge for themselves the possible consequences of a technology, the likelihood that these consequences will occur, their importance, and the combined implications of these various considerations.

Despite an appearance of objectivity, risk assessment is inherently subjective. Rarely will relevant statistical data (such as historical failure rates) be available. When they are, interpretation of such data is still subjective. More often, especially with new technologies, the risks must be estimated by applying engineering judgment to blueprints or to data on related systems and test trials. For the lay person, lacking specialized training and access to data, the decision process will be even more subjective.

This chapter explores some of the intellectual elements in risk assessment that are critical to the nuclear debate. Its basic premises are that both the public and the experts are necessary participants in that debate, that there is a subjective element in all judgments, and that understanding the limitations of judgmental processes and proposed decision-making techniques is crucial to effective hazard management.

COPING INTELLECTUALLY WITH RISK

Decisions about nuclear energy require serious thought and reasoning on the part of experts and nonexperts alike. They require an appreciation of the probabilistic nature of the world and the ability to think intelligently about low-probability (but high-consequence) events. As Weinberg (1976) noted, "we certainly accept on faith that our human intellect is capable of dealing with this new source of energy" (p. 21). Recently, however, the faith of many of us who study human decision processes has been shaken.

Consider probabilistic reasoning. Because of its importance to decision making, a great deal of recent research has been devoted to understanding how people deal with the probabilities of uncertain events. This research has found that intelligent people systematically violate the principles of rational decision making when judging probabilities, making predictions, and otherwise attempting to cope with uncertainty. For example, people fail to recognize randomness when they encounter it; instead they perceive systematic patterns and lawful relationships in situations where none exist. In some situations, they overvalue small samples of data and unreliable data. Yet, when attempting to make forecasts or predictions, they desire too much information, often of the wrong type. Frequently, these difficulties can be traced to the use of judgmental "heuristics," mental strategies (or rules of thumb) that allow people to reduce difficult tasks to simpler judgments (Tversky and Kahneman, 1974). These heuristics are useful guides in some circumstances, but in others they lead to large, persistent biases with serious implications for decision making.[1]

Availability Bias

One heuristic particularly relevant for judgments about risks from nuclear power is "availability." This heuristic involves viewing an event as likely or frequent if it is easy to imagine or to recall instances of it. Generally, instances of frequent events are more easily recalled than instances of infrequent events, and likely occurrences are easier to imagine than unlikely ones. Thus, availability is often an appropriate cue for judging frequency and probability. Availability, however, is also affected by numerous factors unrelated to likelihood. As a result, reliance on it may lead people to exaggerate the probabilities of events that are particularly recent, vivid, or emotionally salient.

Availability helps explain distortions in our perceptions of risk. Consider fears about grizzly bear attacks in our national parks. Although many people are concerned about the dangerousness of grizzlies, the rate of injury is only one per two million visitors and the rate of death is very much lower (Herrero,

1. More extensive discussions of heuristics and biases in probabilistic thinking are available in Kahneman, Slovic, and Tversky (1982), Slovic, Kunreuther, and White (1974), Slovic, Fischhoff, and Lichtenstein (1977), and Tversky and Kahneman (1974).

1970). Sensational media reports contribute to the imaginability of death at the claws of an enraged grizzly, but the media ignore the multitude of uneventful visits. The motion picture *Jaws* has likewise increased the availability (and the perceived likelihood) of shark attacks. Some nuclear power proponents feel that the risks of that technology are exaggerated in the public's eye because of excessive media coverage and association with the vivid, imaginable, memorable dangers of nuclear war. As Zebroski (1975) notes, "fear sells"; the media dwell on potential catastrophes, not on the successful day-to-day operations of power plants.

Availability bias is illustrated in a recent study in which college students and members of the League of Women Voters were asked to judge the annual frequencies of death from each of 41 causes, including diseases, accidents, homicide, suicide, and natural hazards (Lichtenstein et al., 1978). These frequency judgments were greatly in error for many of the causes. Table 5.1 lists the causes whose frequencies were most seriously misjudged. Consistent with availability and considerations, overestimated items tended to be dramatic and sensational. Underestimated items tended to be unspectacular events, which claim one victim at a time and are common in nonfatal form.

Table 5.1 Bias in Judged Frequency of Death

Most Overestimated	Most Underestimated
All accidents	Smallpox vaccination
Motor vehicle accidents	Diabetes
Pregnancy, childbirth, and abortion	Stomach cancer
Tornadoes	Lightning
Flood	Stroke
Botulism	Tuberculosis
All cancer	Asthma
Fire and flames	Emphysema
Venomous bite or sting	
Homicide	

Overconfidence

A particularly pernicious aspect of heuristics is that people are typically very confident in the judgments based upon them. Evidence for this comes, for example, from research by Fischhoff, Slovic, and Lichtenstein (1977) in which people were asked to indicate the odds that they were correct in their judgments about which of two causes of death was more frequent. Odds of 100 to 1 or greater were given often (25 percent of the time). However, about one in eight answers associated with such extreme confidence was wrong (fewer than 1 in 100 should have been wrong if the odds had been appropriate). To take but one example, about 30 percent of the judges gave odds greater than 50 to 1 to the incorrect assertion that homicides are more frequent than suicides. The psychological basis for this unwarranted certainty seems to be people's insensitivity to the tenuousness of the assumptions upon which their judgments are based (in this case, the validity of the availability heuristic). The danger from such overconfidence is that we may not realize how little we know and how much additional information we need about the various problems and risks we face.

Overconfidence manifests itself in other ways as well. For example, when experts estimate failure rates or other uncertain quantities they often set upper and lower bounds such that there is a 98 percent chance that the true value lies between them. Experiments with diverse groups of people (including experts) making many different kinds of judgments have shown that, rather than 2 percent of true values falling outside the 98 percent confidence bounds, 20–50 percent do so (Lichtenstein, Fischhoff, and Phillips, 1982). This shows that people think that they can specify such quantities with much greater precision than is actually the case.

Additional evidence of overconfidence in experts is provided by Hynes and Vanmarcke (1976), who asked seven "internationally known" geotechnical engineers to predict the height of an embankment that would cause a clay foundation to fail and to specify confidence bounds around this estimate that were wide enough to have a 50 percent chance of enclosing the true failure height. The bounds specified by these experts were too narrow. None of them enclosed the true failure height. The multimillion dollar Reactor Safety Study (Nuclear Regulatory Commission, 1975), in assessing the probability of a core melt in a nuclear reactor, used the very procedure for setting confidence bounds that has been found in experiments to produce the highest degree of overconfidence. The Committee on Government Operations (U.S. Government, 1976) has attributed the 1976 collapse of the Teton Dam to the unwarranted confidence of engineers who were absolutely certain they had solved the many serious problems that arose during construction. Indeed, in routine practice, failure probabilities are not even calculated for new dams even though about one in three hundred fails when its reservoir is first filled.

Desire for Certainty

Every technology is a gamble of sorts and, like other gambles, its attractiveness depends on the probability and size of its possible gains and losses. Both scientific experiments (see Lichtenstein and Slovic, 1973) and casual observations show that people have difficulty thinking about and resolving the risk/benefit conflicts even in simple gambles (such as choosing which game they would rather play: a gamble with favorable chances of winning a modest amount or a gamble with less favorable odds but a larger possible payoff).

One way to reduce the anxiety generated by uncertainty is to deny the uncertainty. The denial resulting from this anxiety-reducing search for certainty constitutes another source of overconfidence, in addition to those described earlier. Denial is illustrated by many people exposed to natural hazards who view their world as either perfectly safe or predictable enough to preclude worry. Thus some flood victims interviewed by Kates (1962) flatly denied that floods could ever recur in their areas. Some thought (incorrectly) that new dams and reservoirs in the area could contain all potential floods, while others attributed previous floods to freak circumstances unlikely to recur. Denial, of course, has its limits. Many people feel that they cannot ignore the risks of nuclear power. For these people, the search for certainty is best satisfied by outlawing the risky technology.

Scientists and policy makers who point out the gambles involved in societal decisions are often resented for the anxiety they provoke. Borch (1968) noted that corporate managers get annoyed with consultants who give them the probabilities of possible future events instead of telling them exactly what will happen. Just before hearing a blue-ribbon panel of scientists report that they were 95 percent certain that cyclamates do not cause cancer, Food and Drug Administration commissioner Alexander Schmidt said, "I'm looking for a clean bill of health, not a wishy-washy, iffy answer on cyclamates" (*Eugene Register Guard,* January 14, 1976). Senator Muskie has called for "one-armed" scientists, who do not respond "on the one hand, the evidence is so, but on the other hand . . ." when asked about the health effects of pollutants (David, 1975).

The search for certainty is legitimate if it is done consciously, if residual uncertainties are acknowledged rather than ignored, and if people realize the costs. If extreme certainty is sought, those costs are likely to be high. Eliminating the uncertainty may mean eliminating the technology and foregoing its benefits. Often, some risk is inevitable. Efforts to eliminate it may only alter its form. We must choose, for example, between the vicissitudes of nature on an unprotected flood plain and the less probable, but potentially more catastrophic, hazards associated with dams and levees.

Perseverence of Beliefs

The difficulties of facing life as a gamble contribute to the polarization of opinion about nuclear power. Some people view it as extraordinarily safe, while others view it as a catastrophe in the making. It would be comforting to believe that these divergent beliefs would converge toward one "appropriate" view as new evidence was presented. Unfortunately, this is not likely to be the case. As noted earlier in our discussion of availability, risk perception is derived in part from fundamental ways of thinking that lead people to rely on fallible indicators such as memorability. Furthermore, a great deal of research indicates that people's beliefs change slowly and are extraordinarily persistent in the face of contradictory evidence (Ross, 1977). Once formed, initial impressions tend to structure and distort the way in which subsequent evidence is interpreted. New evidence appears reliable and informative if it is consistent with one's initial belief; contradictory evidence is dismissed as unreliable, erroneous, or unrepresentative. Ross (1977) concluded his review of this phenomenon as follows:

> Erroneous impressions, theories, or data processing strategies, therefore, may not be changed through mere exposure to samples of new evidence. It is not contended, of course, that new evidence can *never* produce change—only that new evidence will produce *less* change than would be demanded by any logical or rational information-processing model. [p. 210]

The "Wisdom" of Hindsight

Technologists, policy makers, and regulators must constantly consider how their decisions will be judged in hindsight by history, by the public, and by courts of law. Their decisions will be judged harshly if it appears that they failed to anticipate important, foreseeable difficulties. Psychological research indicates that, in hindsight (that is, knowing how things actually turned out), people consistently exaggerate what could have been anticipated in foresight (Fischhoff, 1975a, 1975b). They not only tend to view events that happened as having been inevitable, but also believe (often incorrectly) that those events appeared to be "relatively inevitable" before they happened and that "others should have known they were going to occur." They also misremember their own predictions, exaggerating in hindsight what they themselves knew in foresight (Fischhoff, 1977a; Fischhoff and Beyth, 1975).

An extreme response to this form of bias is to claim that one's critics are always guilty of capitalizing unfairly on hindsight knowledge. This would not be entirely justified, for although the bias is pervasive, there is still some relation between what was and what seems to have been anticipatable. Fair evaluation requires a method for improving hindsight. One bit of advice we can give to decision makers is to leave a clear record of what they knew and the uncertainties surrounding their actions. Critics of these decision makers might improve the acuity of their hindsight by attempting to state how the event might have turned out otherwise (Slovic and Fischhoff, 1977) or by seeking the opinions of persons not already affected by knowledge of the outcome.

Forecasting Public Response toward Nuclear Power

Given this research into the ways in which people generally view risks and uncertainties, can we predict how they will respond to nuclear power? Probably not very well. Depending on how nuclear risks are presented by the media, in public debates, and in private discussions and on whether or not there are major accidents, near misses, or energy shortages, nuclear power may come to be viewed as increasingly safe or increasingly dangerous.

Implications of the Availability Heuristic. It is easy to see how accidents, near misses, or even minor problems, coupled with the attention the news media give such events, would increase the perception of risk from nuclear power. But a more subtle and disturbing implication of the availability heuristic is that any discussion of low-probability hazards, regardless of its content, will increase the memorability and imaginability of those hazards and, hence, increase their perceived risks. This poses a major barrier to open, objective discussions of nuclear safety. Consider an engineer demonstrating the safety of waste disposal in a salt bed by pointing out the improbability of the various ways radioactivity could be released (see figure 5.1). Rather than reassuring the audience, the presentation might frighten them with the thought that there are more things than they had realized that could go wrong.

Availability magnifies fears of nuclear power by blurring the distinction between what is remotely possible and what is probable. As one nuclear proponent lamented, "When laymen discuss what *might* happen, they sometimes don't even bother to include the 'might' " (B. L. Cohen, 1974, p. 36). Another ana-

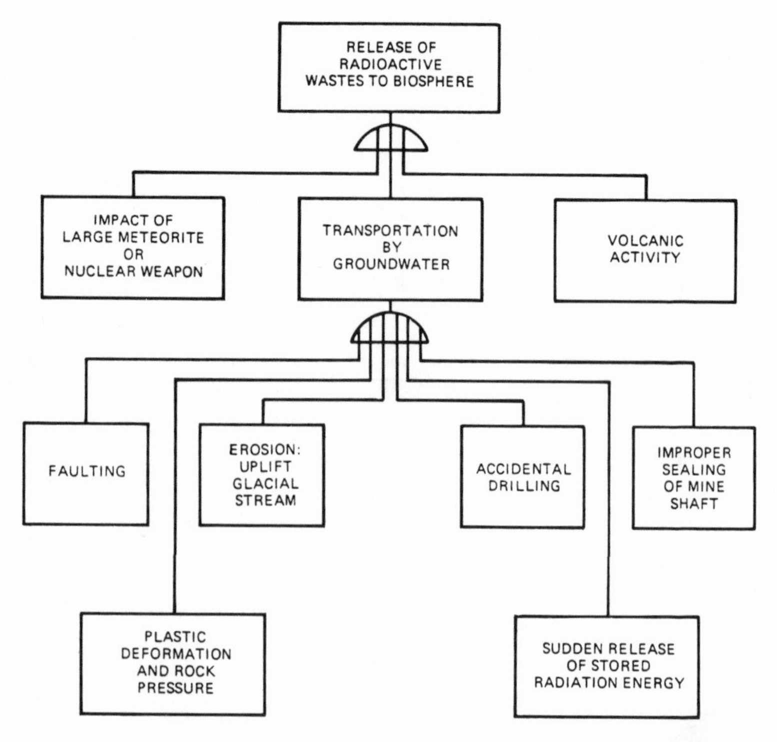

Figure 5.1. Fault tree indicating the possible ways that radioactivity could be released from wastes deposited in bedded salt (after closure of the repository) (from McGrath, 1974)

lyst has elaborated a similar theme in the misinterpretation of "worse case" scenarios:

It often has made little difference how bizarre or improbable the assumption in such an analysis was, since one had only to show that some undesirable effect could occur at a probability level greater than zero. Opponents of a proposed operation could destroy it simply by exercising their imaginations to dream up a set of conditions which, although they might admittedly be extremely improbable, could lead to some undesirable results. With such attitudes prevalent, planning a given nuclear operation becomes somewhat perilous since it requires predicting the extent to which the adversaries can employ their imagination (J. J. Cohen, 1972, p. 55).

Whereas the preceding discussion helps to clarify the source of the perception gap between pronuclear experts and their lay opponents, it does not point unambiguously to one side or the other as having the most accurate appraisal of the overall risks from nuclear power. Although memorability and imaginability are capable of enhancing public fears, inability to imagine all the possible ways that a system could fail might produce a false sense of security among technical experts. As a result, the identification of judgmental difficulties does not, in itself, afford an external criterion for closing the perception gap.

Qualitative Factors. Opponents of nuclear power appear to be responding not just to the probability of a mishap, but also to a number of frightening qualitative perceptions of the harm it might produce. Some insight into these qualitative factors may be found in a study by Fischhoff, Slovic, Lichtenstein, Read, and Combs (1978), in which seventy-six members of the Eugene, Oregon, League of Women Voters rated the risks from thirty activities and technologies on nine qualitative scales: voluntariness, familiarity, controllability, potential for catastrophe (multiple fatalities), immediacy of consequences, degree to which the risks are known to the public and to scientists, extent to which the risks are common (as opposed to dread), and lethality (the likelihood that a mishap would prove fatal). Participants also rated the total risks and benefits accruing to society from each activity or technology, as well as how acceptable those risks were.

One of the most interesting findings of that study was the evaluation of nuclear power. For one, the benefits of nuclear power were not appreciated, being judged lower than those of home appliances, bicycles, and general aviation. Perhaps this is because nuclear power is seen merely as a supplement to other, essentially adequate, sources of energy. Second, its risks were seen as extremely high. Only automobile accidents, which take about 50,000 lives each year, were viewed as comparably risky. Third, its current level of risk was judged as unacceptably high. Participants in this study wanted nuclear power to be far safer than they now perceive it to be. The frightening character of nuclear power emerged clearly in the rating scales. Figures 5.2a and 5.2b show its unique risk profile. Nuclear power was rated at or near the extreme on all of the characteristics associated with high risk: involuntariness, uncontrollability, dread, lethality, etc. These figures also contrast nuclear power with two ostensibly similar technologies, X rays and nonnuclear

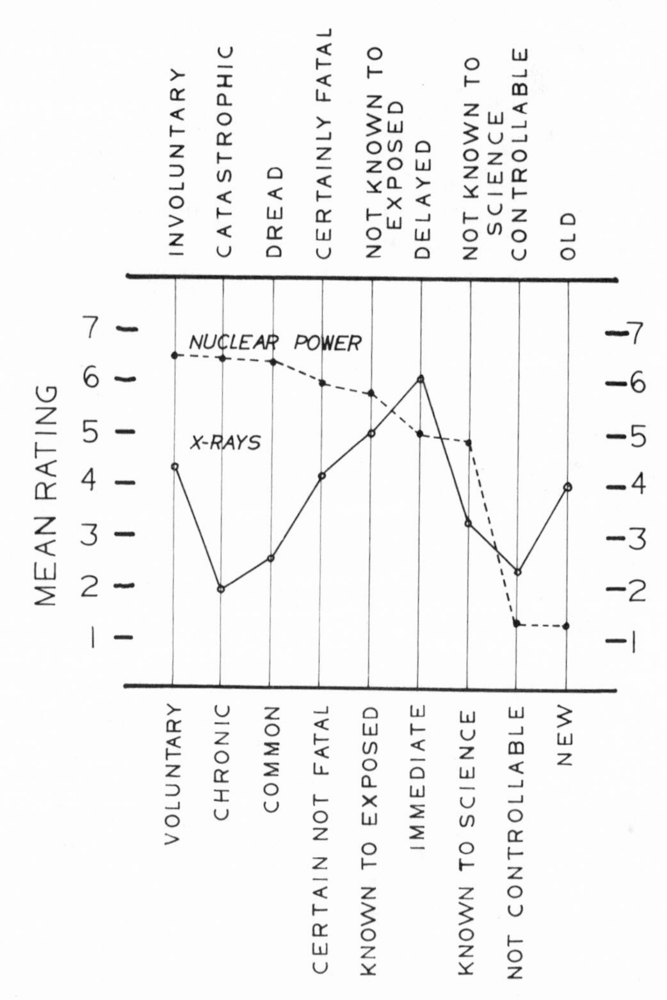

Figure 5.2a. Comparison between nuclear power and X rays on nine risk characteristics (from Fischhoff et al., 1978)

electric power. Although both X rays and nuclear power involve radiation, nuclear power was judged much more catastrophic and dread. The comparison in figure 5.2b shows that, where risk is concerned, nuclear power is not seen as just another form of energy.

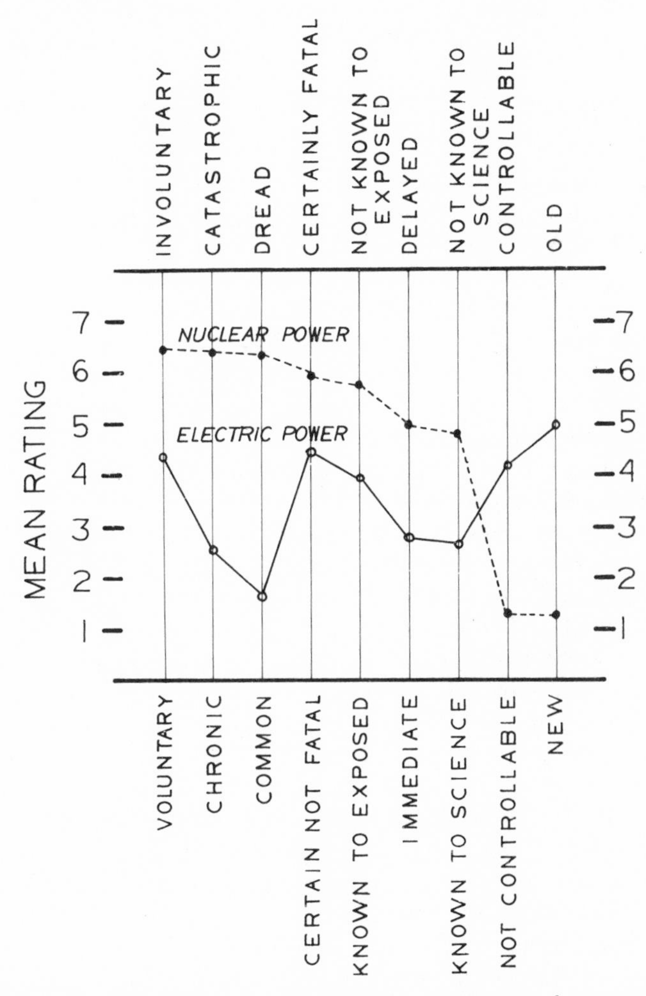

Figure 5.2b. Comparison between nuclear power and nonnuclear electric power on nine risk characteristics (from Fischhoff et al., 1978)

Nuclear power was in, fact, rated higher on "dread" than any of the other items studied by Fischhoff et al. This may stem from the association of nuclear power with nuclear weapons and from fear of radiation's invisible and permanent bodily contamination that causes genetic damage and cancer (Lifton, 1976; Pahner, 1975). Subsequent studies using students, business people, and professional risk experts as participants have revealed remarkable agreement in judgments of the qualitative characteristics of the risks encountered with these thirty technologies.

Pathways toward Acceptance. With all this working against it, how could nuclear power ever gain acceptance? Public response to X rays and nerve gas provides some clues. Widespread acceptance of X rays suggests that a radiation technology can be tolerated once its use becomes familiar, its benefits clear, and its practitioners trusted. Although nuclear power might someday attain the low "dread" level of X rays, however, its perceived potential for castastrophic accidents seems less likely to change. Whether a continued high score on that one characteristic would render it permanently unacceptable is unclear.

Nerve gas may provide an enlightening case study. Few human creations could be more dread or more potentially catastrophic than this deadly substance. When, in December of 1969, the army decided to transfer nerve gas from Okinawa to the Umatilla Army Depot in Hermiston, Oregon, citizens of Oregon were outraged—except those in Hermiston. Whereas public opinion around the state was more than 90 percent opposed, residents of Hermiston were 95 percent in favor of the transfer, despite the warning that the fuses on the gas bombs deteriorate with age, but that the gas does not (*Eugene Register Guard,* December 18, 1969; January 11, 1970). Several factors seem to have been crucial to Hermiston's acceptance of nerve gas. For one, munitions and toxic chemicals had been stored safely there since 1941, so the record was good and presence of the hazard was familiar. Second, there were clear economic benefits to the community from continued storage of hazardous substances at the depot, in addition to the satisfaction of doing something patriotic for the country. Finally, the responsible agency, the U.S. Army, was respected and trusted.

These examples illustrate the slow path through which nuclear power might gain acceptance. It requires an incontrovertible long-term safety record, a responsible agency that is respected and trusted, and a clear appreciation of benefit. However, since people are generally willing to accept increased risks in exchange for increased benefits (Starr, 1969; Fischhoff et al., 1978), a quicker path to acceptance might be forged by a severe energy shortage. Brownouts, rationing, or worse would undoubtedly enhance the perceived benefits from nuclear power and increase society's tolerance of its risks. A recent example of this process is the oil crisis of 1973–74, which broke the resistance to offshore drilling, the Alaska pipeline, and shale oil development, all of which had previously been delayed because of environmental concerns. Such crisis-induced acceptance of nuclear power may, however, produce anxiety, stress, and conflict in a population forced to tolerate what it perceives as great risk because of its addiction to the benefits of electricity.

Resistance to Change. One likely possibility is that people will maintain their present positions, pro or con. This would be consistent with the research on perseverance of beliefs showing that rather than modifying existing beliefs, new evidence tends to be interpreted in a way that confirms them. Thus the accident at Three Mile Island proved the possibility of a catastrophic meltdown to some, whereas to others it demonstrated the reliability of the multiple containment systems.

AIDING THE DECISION-MAKING PROCESS

The intellectual limitations described above portend continued and severe conflict over the safety and desirability of nuclear energy. The prospect of poor decisions with extreme costs to society seems all too likely. Sinsheimer (1971) observed that the human brain has evolved to cope with real, immediate, and concrete problems and thus lacks the proper framework with which to encompass other sorts of phenomena. People have only recently faced decisions such as those involving nuclear energy. Following Sinsheimer's reasoning, it might be argued that we have not had the opportunity to evolve the intellect needed to deal with uncertainties of this nature. We are essentially trial-and-error learners, in an age when errors are increasingly costly.

But there may be some positive steps we can take to minimize the consequences of these limitations. Besides being less confident in our intellect, we can attempt to develop procedures for combatting the biases to which risk assessments are susceptible. The simplest "procedure" is to be wary of bias and hope that alertness will suffice. Alternatively, we may employ several methods for making the same judgment in the hope that their respective biases will be detected or perhaps balance one another. A third possibility is to restructure judgment tasks, perhaps by decomposing them into simpler judgments that can be made with less bias. The use of fault trees is a widely used decomposition method for estimating risks that we shall discuss later in this section.

The same technological bent that has created so many new hazards has also created methods designed to help make decisions about these hazards. Cost-benefit analysis and decision analysis are leading members of this genre. Like the technologies they are meant to evaluate, these analytic techniques have both potential benefits and inherent limitations. They can improve the decision-making process and its sensitivity to public desires, but only if the public understands the techniques and their limitations, monitors the way that analyses are done, and makes certain that their conclusions are heeded (or ignored, as suitable). To this end, the present section also describes some of the more important approaches and techniques for making decisions, the problems encountered in applying them, and the ways in which specific techniques may be led astray and produce erroneous results.

Estimating Risks

A frequently used aid for assessing and communicating the risks of a complex system is a fault tree. Construction of a fault tree begins by listing all important pathways to failure, then listing all possible pathways to these pathways and so on, as shown in figure 5.1. A fault tree representing the risks of an automobile failing to start appears in figure 5.3. The first level lists major systems problems such as battery failure; the next level traces these global failures to sources like loose terminals or weak charge; the lowest level provides even more detail. When the desired degree of detail is obtained, the experts assign probabilities to each of the component pathways (relying on judgment or available data) and then combine these to provide an overall failure rate.[2] The importance of fault-tree analysis is demonstrated by its role as the primary methodological tool in the Reactor Safety Study, which assessed the probability of a catastrophic loss-of-coolant accident in a nuclear power reactor (Nuclear Regulatory Commission, 1975).

Errors of Omission. Fault-tree analysis has been attacked by critics who question whether it is valid enough to be used as a basis for decisions of great consequence (see Bryan, 1974; Fischhoff, 1977a; Primack, 1975). One major danger in designing a fault tree is leaving things out and, thereby, underestimating the true risk. The car-won't-start tree would be seriously deficient if it failed to include problems with the seat-belt system (for 1974 models) or vandalism.

Several kinds of pathways seem particularly prone to omission. One type involves human error or sabotage. Can we ever be certain that we have enumerated all of the important and imaginative ways in which we, the people (as opposed to they, the machines), can create trouble? Consider the fire in the large nuclear power reactor in Browns Ferry, Alabama, which was caused by a technician checking for an air leak with a candle, in direct violation of standard operating procedures (Comey, 1975). The fire got out of control, in part, because plant personnel were slow to sound alarms and begin the re-

2. Fault trees start with a particular undesired final event (a failure of the system) and work backward to identify the component failures needed to initiate that event. A related method uses event trees. These start from a particular initiating event (such as an earthquake in a waste storage area) and project all possible outcomes of that event.

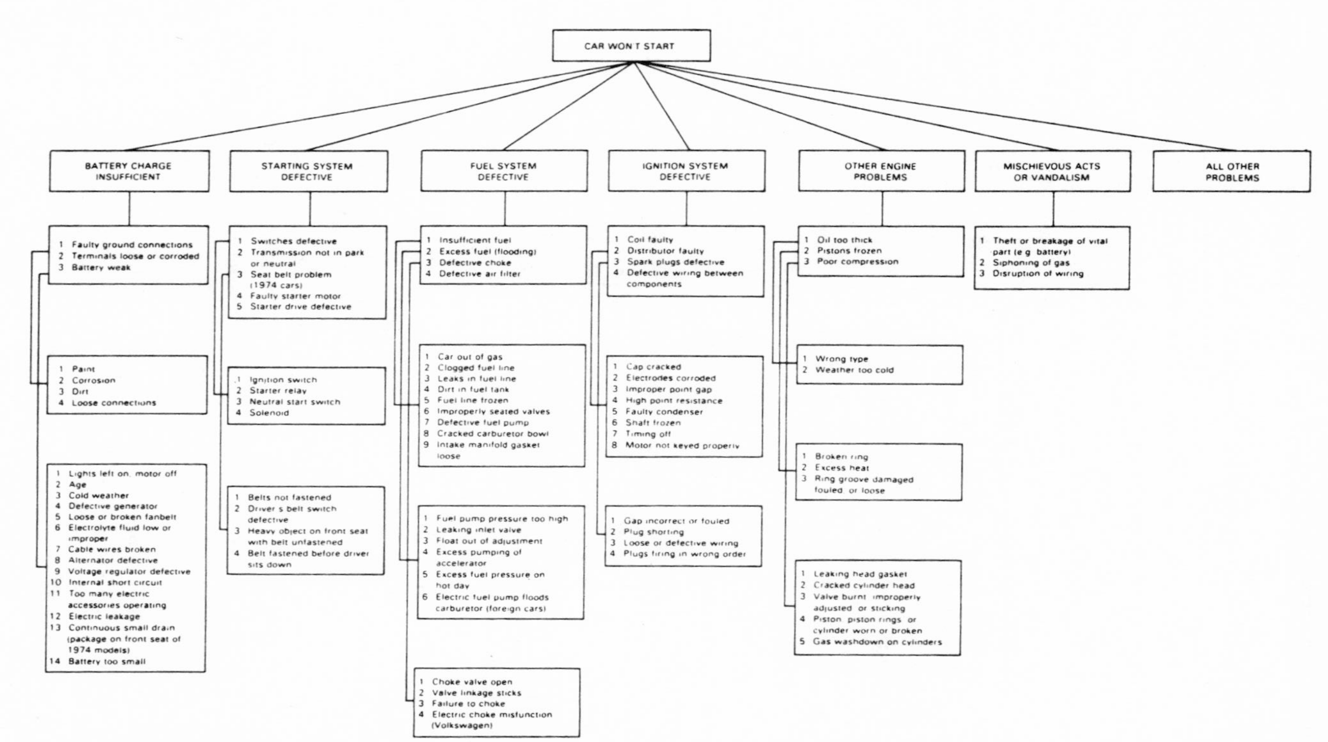

Figure 5.3. A fault tree indicating the ways in which an automobile might fail to start (from Fischhoff, Slovic, and Lichtenstein, 1978)

actor shut-down. Disaster was averted finally when plant personnel managed to jury-rig pumps normally used to drive control rods into the reactor to, instead, pump water to cool the reactor core. Identifying such possibilities for human error (or ingenuity) obviously poses a difficult challenge for risk analysts (Nuclear Regulatory Commission, 1978).

A second source of omissions is failure to consider unanticipated changes in the world in which the technology functions (Coates, 1976; Hall, 1975). Risk assessments always assume some constancies in the external environment. At times, these assumptions may be both unrecognized and questionable. For example, nuclear power plant design assumes implicitly the continued availability of properly trained personnel. Even though a tree's designers might not realize that they are making this assumption, it is possible to imagine a future world in which such individuals are in short supply.

Omissions may also result from failing to see how the system functions as a whole. For example, the rupture of a liquid natural gas storage tank in Cleveland in 1944 resulted in 128 deaths, largely because no one had realized the need for a dike to contain spillage (Katz and West, 1975). The DC-10 failed repeatedly in its initial flights because none of its designers realized that decompression of the cargo compartment would destroy vital parts of the plane's control system (Hohenemser, 1975). Green and Bourne (1972, p. 547) caution us not to forget that backup systems may not function when needed because they are undergoing routine maintenance and testing or because they have been damaged by the testing process.

Another omission in this category is provided by a National Academy of Sciences study of the effects of thermonuclear war. The academy's panel decided that the anticipated reduction of the earth's ozone shield would not imperil the survivors' food supply because many crops could survive the increased ultraviolet radiation. The study failed to point out, however, that increased radiation would make it virtually impossible to work in the fields to raise those crops. "How was this overlooked? Because . . . it fell between the chinks of the expert panels. The botanists who considered the effects of ultraviolet radiation on plants didn't think to worry about the workers" (Boffey, 1975, p. 250).

Common-Mode Failures. Another sort of error in the use of fault trees and one that the Reactor Safety Study took great pains to avoid is miscalculating what are called "common-mode failures." To insure greater safety, many technological systems are built with a great deal of redundancy. Should one crucial part fail, there are others designed either to do the same job or to limit the resulting damage. Since the probability of each individual part failing is very small, the probability of all failing, thereby creating a major disaster, would seem to be extremely small. This reasoning is valid only if the various components are independent, that is, if what causes one part to fail will not automatically cause the others to fail. Common-mode failure occurs when the independence assumption does not hold. As an example, the discovery that a set of pipes in several nuclear power plants were all made from the same batch of defective steel (*Eugene Register Guard,* October 13, 1974) suggests that the simultaneous failure of several such pipes is not inconceivable. At Browns Ferry, the same fire that caused the core to overheat also damaged the electrical system needed to shut the plant down. Constructing a tree that takes proper account of all such contingencies may be very difficult (Nuclear Regulatory Commission, 1978).

Presentation Biases. Fault trees are tools for the communication as well as for the estimation of risks. The expert who has completed an analysis of risks, with whatever success, must present the results to other experts or to the public. Doing so involves making a number of discretionary decisions. For example, with the car starting fault tree (figure 5.3), the presenter must decide how much detail to provide for each branch, which minor pathways to lump into the "all other problems" category, and just how to categorize various items.

We have earlier studied how such discretionary aspects of fault-tree presentation affect people's perceptions of the risks they embody (Fischhoff, Slovic, and Lichtenstein, 1978). Our results indicate that the decision to put some difficulties in an "all other problems" category can have a major biasing effect. People are quite insensitive to how much has been left out of a fault tree. Deleting branches responsible for about half of all automobile starting failures only produced a 7 percent increase in people's estimates of what was missing. Professional automobile mechanics were about as insensitive as nonexperts. Apparently, what was out of sight was also out of mind. The fault-tree presenter who, deliberately or inadvertently, fails to mention a branch (thereby implicitly or explicitly assigning it to "all other problems") may remove it completely from consideration. We also found that the perceived importance of a set of problems can be substantially increased by presenting it as two smaller problem categories rather than as one category.

That subtle differences in how risks are presented can have large effects on how they are perceived suggests that people attempting to communicate information about risks have considerable ability to manipulate others' perceptions without making any overt misrepresentations. Indeed, since these effects are not widely known, people may inadvertently be manipulating their own perceptions by decisions they make about how to organize their knowledge.

As with the other research we have described, studies of presentation biases have two lessons. One is to be wary, realizing that judgment is fallible and readily influenced by irrelevant factors. The second is that these effects might be counteracted by adopting a variety of perspectives. Ask yourself: What is left out? How would this problem look if the categories were rearranged? Is the presenter interested in manipulating my perceptions and, if so, what strategies might make that possible?

Reaching a Decision

Muddling Through. After risks have been assessed, some decision must be made. By far the most common approach toward setting risk policies, nuclear or otherwise, is "muddling through," making somewhat arbitrary initial decisions and then letting them be molded into generally accepted standards by the pressure of political and economic forces. While this process may employ analytic arguments, it is essentially nonanalytic. It relies upon the internal structure of participating organizations, their interaction with one another, and the varied feedback provided by their environment to produce satisfactory decisions. The building blocks of this approach are mechanisms that are familiar and accepted, even if not entirely understood. This approach does not lend itself to producing and defending acceptable risk criteria on the basis of specific analytic techniques, although it does allow such analyses to be presented as evidence.

Comparative Analyses. One major form of input into the process of muddling through comes from various forms of comparative analysis. Comparative procedures are used to determine the acceptable level of risk for a given hazard (such as nuclear power) by reference to the level of safety tolerated from other hazards, either natural or technological. For example, the allowable radiation from a particular segment of the nuclear fuel cycle might be set equal to natural background radiation or to a fraction of tolerated radiation from medical exposures. Workers in the nuclear industry might be expected to tolerate the same level of risk borne by workers in other energy-producing industries.

Comparative analysis has several attractive features. It avoids the difficult and controversial task of converting diverse risks into a common monetary unit (like dollars per life lost or per case of sterilization or per day of suffering). It presents issues in a mode that is probably quite compatible with natural thought processes. Among other things, this mode may avoid any direct numerical reference to very small probabilities, for which people have little or no intuitive feeling.

A more elaborate form of comparative analysis, incorporating benefits as well as risks, is the "revealed preference" approach advocated by Starr (1969). This approach is based on the assumption that, by trial and error, society has arrived at nearly optimal balance between the risks and benefits associated with any activity. If this is the case, then historical data can be used to reveal acceptable trade-offs of risks and benefits. Acceptable risk for a new technology is assumed to be the level of safety associated with ongoing activities having similar benefit to society.

From this approach, Starr derived what may be regarded as "laws of acceptable risk." Two of these laws are that greater risks are acceptable for more beneficial activities and that the public accepts much greater risks from voluntary activities (such as skiing) than from involuntary activities (such as ingesting food preservatives), even if they provide similar levels of benefit. Thus, according to Starr's analysis, acceptable risk is determined by two factors, benefit and voluntariness.

Although the method of revealed preference is based upon an intuitively compelling logic, it has several drawbacks. It assumes that past behavior is a valid predictor of present preferences, perhaps a dubious assumption in a world where values may change quite rapidly. It is politically conservative in that it enshrines current economic and social arrangements. It makes strong and not always supported assumptions about the rationality of people's decision making in the marketplace and about the freedom of choice that the marketplace provides. It may underweigh risks to which the market responds sluggishly, such as those, like most carcinogens, that reveal themselves slowly. Finally, it is no simple matter to develop the measures of risks and benefits needed for its implementation (Otway and Cohen, 1975).

Cost-Benefit Analysis. Cost-benefit analysis is a method of quantifying, in terms of dollars, the expected gains and losses from a proposed action. If the calculated benefits from a project are greater than its costs, the action is recommended. When risks to life and health represent an important component of the costs, the term "risk-benefit analysis" is used.

The expected cost of a project is determined by enumerating all aversive consequences that might arise from its implementation, assessing the probability that each will occur, and estimating the cost or loss to society should each occur. Next, the expected loss from each possible consequence is calculated by multiplying the amount of the loss by the probability that it will be incurred. The expected loss of the entire project is computed by summing the expected losses associated with the various possible consequences. An analogous procedure produces an estimate of the expected benefits.

Performing a full-dress analysis assumes, among other things, that all significant consequences can be enumerated in advance; that meaningful probability, cost, and benefit judgments can all be reduced to dollar equivalents; that people really know how they value consequences today and how they will value them in the future; and that people want, or should want, to maximize the difference between expected benefits and losses (Fischhoff, 1977b).

Decision Analysis. Cost-benefit analyses typically produce a yes-no decision about one particular project; its costs either do or do not outweigh its benefits. Often, however, we are faced with an array of alternative actions from which we must choose the best. In addition, cost-benefit analysis only considers consequences that can be expressed in dollars. For many amenities and disamenities (aesthetic improvement or degradation, for example) monetary values cannot be assigned.

Recently, a technique called decision analysis has been developed to handle situations with multiple alternatives and varied consequences. Decision analysis is hailed by some as the general method of choice for coping with risky decisions (Howard, 1968, 1975; Keeney and Raiffa, 1976; Raiffa, 1968). It combines sophisticated modeling of decision problems (that is, the critical options, events, and consequences) with a theory specifying how to deal rationally with uncertainty and the inevitable subjectivity of decision makers' preferences and values. Decision analysis has been applied to problems involving hurricane modification (Howard, Matheson, and North, 1972), the selection of experiments for a Mars space mission (Matheson and Roths, 1967), the decision to undergo coronary artery surgery (Pauker, 1976), and the desirability of nuclear power plants (Matheson et al., 1968; Barrager, Judd, and North, 1976).

In some ways, decision analysis is a generalized form of cost-benefit analysis. After identifying feasible courses of action, possible outcomes are specified and the probabilities and values of those outcomes are determined. The alternative of choice is the one with the highest expected value, the greatest preponderance of expected benefits over expected costs. As with cost-benefit analysis, fault trees could be used to assess probabilities and dollars could be used as the common unit of value. In fact, one could do a decision analysis of just one alternative (or, rather, of the two alternatives "act" and "don't act"), reducing the decision analysis to a slightly modified cost-benefit analysis. More often, however, the decision analyst will choose a common unit of value, usually called "utility," with no necessary relation to dollars. This makes it easier to include consequences such as increased anxiety and aesthetic degradation, which have no convenient dollar equivalent.

Decision analysis is characterized not only by its conceptual framework, but also by some of the techniques it uses. Much emphasis is placed on what might be called "dynamic" modeling of the decision problem, looking at how possible actions interact with world events. For example, a decision model might include a sequence like "build plant according to Plan B—*federal emission standards change*—retool plant—*litigation holds up operating permit*—plant starts up 6 months late." Probabilities are assigned to the world events (underlined above) and a value is assigned to the end state, a plant in operation after this course of events.

Decision analysts acknowledge that many (or even all) of the probabilities and values they use are not well defined. They handle this problem with "sensitivity analysis," a technique which leads them to ask such questions as "How much of a difference would it make in the final decision if this probability were off by a factor of 10?" If, when each probability and the value assigned each consequence is varied widely enough to encompass whatever degree of uncertainty exists, the same action is still recommended, one's confidence in the decision is greatly increased. If changes in these values produce different decisions, one should be cautious and collect more data on those values whose fluctuations most influence the decision.

Decision Analysis of Waste Disposal Options. Although a full-scale decision analysis of nuclear waste options is beyond the scope of this paper, a brief outline of how such an analysis might proceed may be instructive.

Let us assume that nuclear power is generating high-level radioactive wastes and a decision has been made to dispose of these wastes permanently. The decision of interest concerns the optimal means of permanent disposal.

Many options are possible, including storage in geologic formations, the sea bed, or polar ice and even extraterrestrial disposal. Each option has several suboptions (for example, different geologic formations). An analysis of risks and benefits needs to be done for each option and suboption.

We shall focus here on one aspect of one option: some of the social costs associated with geological disposal in a cavity within a salt bed. As an additional restriction, even though these social costs are incurred both while the wastes are deposited and during the long time period after the repository is sealed, we shall consider only the latter period.

Figure 5.4 depicts a model for calculating the social costs of disposal in a salt bed after closure. First, all hazards that could trigger a radioactive release (and resultant public loss) are catalogued. The magnitude of loss will depend on the spatial and temporal distribution of the release. Hence, a dispersion model integrating geologic and demographic information is needed to predict human exposure and property contamination. A human implications model is also needed to specify the expected number of deaths and illnesses and the genetic and property damage resulting from this exposure. Finally, all these various damages must be converted to a measure of expected loss that can be integrated with the risks and benefits derived from the other segments of the analysis.

Figure 5.5 outlines the procedures for calculating social loss from salt-bed mishaps in greater detail. First, probabilities must be assessed for each possible category of radioactive release. For some categories, there are relevant historical data from which probabilities can be derived. This is the case for meteorite impacts. Blake (1968) has calculated that the proba-

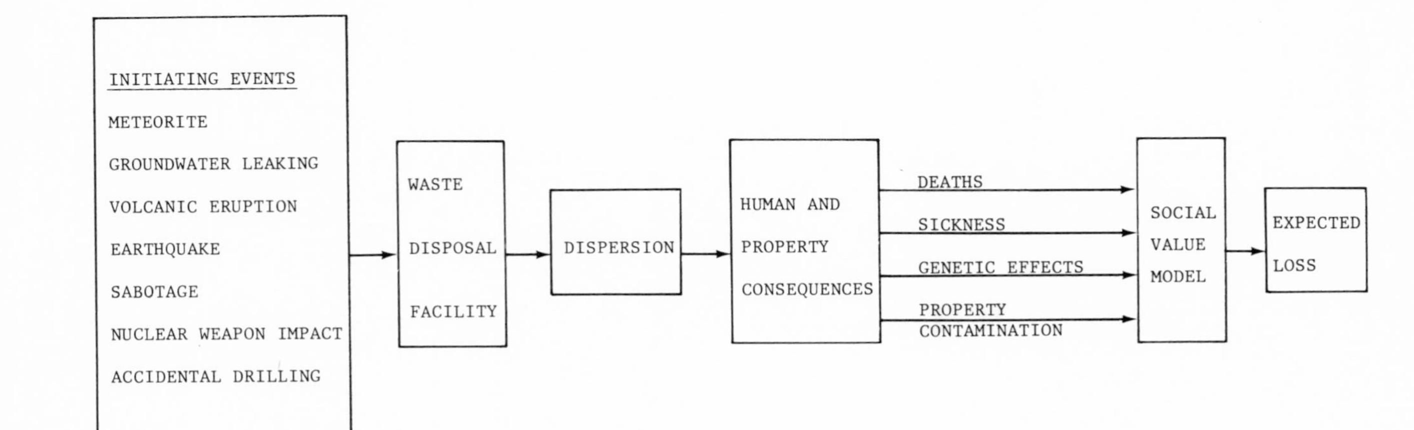

Figure 5.4. A model for calculating the social costs from waste disposal in bedded salt

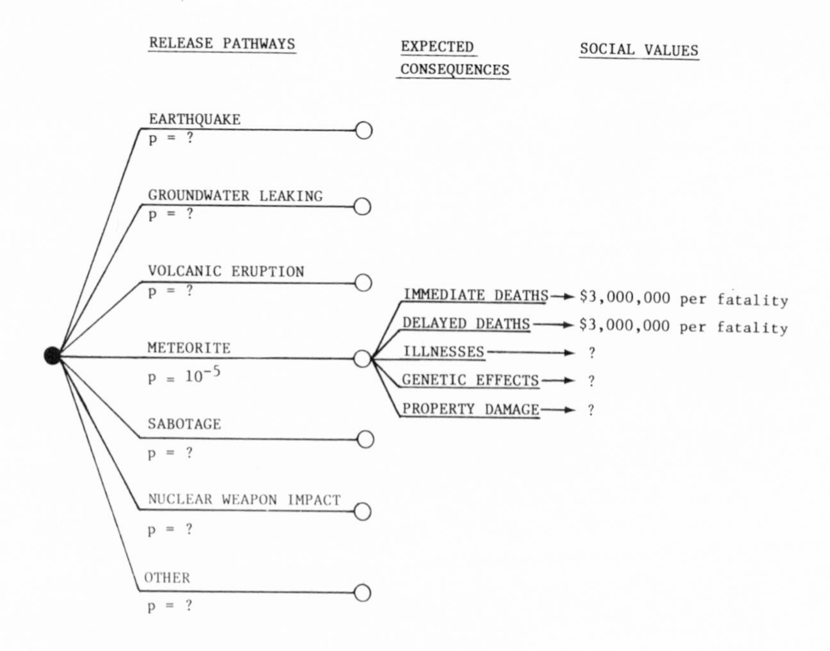

Figure 5.5. Human and property consequences of radiation release from salt bed disposal site

bility of a meteorite making a crater with a width of 200–300 meters in a land storage surface of 100 sq. km. over a period of 100,000 years is about 10^{-5}. For other categories, such as sabotage, we have little recourse but to use probability assessments based on expert judgment, perhaps aided by fault trees or other formal models. The dispersion and human implication models would then be used to calculate expected death and illness for each of the seven possible branches of radiation release.

Finally, a dollar or utility value would have to be assigned to each consequence. For example, one might assign a social cost of $3 million for each death. In that case, if we expected 20 immediate and 20 delayed fatalities from a meteorite impact, each valued at $3 million, the expected loss from this segment of the analysis would be $\$3{,}000{,}000 \times 40 \times 10^{-5}$ which equals \$1200.[3] In similar fashion, the expected loss from illness, genetic effects, and other factors would be computed for meteorite impact and all the other consequences of the remaining branches. Combining the expected losses from the seven branches would produce an overall expected social cost of salt-bed disposal. This value could then be combined with similarly derived figures for economic losses and social and economic benefits from salt-bed disposal. The result, representing the overall expected value of salt-bed disposal, could then be compared with the expected values of other waste disposal options.

Although this analysis is crude and inadequate, it does serve to highlight the following basic questions that will be raised in connection with any thorough decision analysis.

Are the estimates accurate enough for decision making? Earlier in this paper, we discussed the many biases that can affect probability assessments. Certainly all possible steps should be taken to minimize such biases in a decision analysis. For some problems, however, no effective debiasing procedure is known; for others, the presence of bias may not even be recognized. In addition, the analysis must also accommodate disagreement among experts, regarding both probabilities and values assigned to consequences. Sensitivity analysis deals with these problems by using ranges of opinion instead of single estimates. However, unless some alternative stands out as best even when widely disparate estimates are used, decision analysis will not tell us what to do. One might argue that, in such cases, we simply can't know what course of action is best.

Some critics would argue that no estimates are good enough when the events of importance are extremely rare. For example, according to Holdren (1976),

> the expert community is divided about the conceivable realism of probability estimates in the range of one in ten thousand to one in one

3. We could have worked through this calculation with utilities, but the (dis)utility assigned to death would have meant little intuitively without knowing what values were assigned to other consequences. That is, one utility unit per death could be a lot or a little depending on the utility assigned to each illness.

billion per reactor year. I am among those who believe it to be impossible *in principle* to support numbers as small as these with convincing theoretical arguments (that is, in the absence of operating experience in the range of 10,000 reactor years or more), even ignoring the crucial possibility of malevolence. The reason I hold this view is straightforward: nuclear power systems are so complex that the probability the safety analysis contains serious errors (for example, that it omits failure modes more important than those included) is so big as to render meaningless the tiny computed probability of accident. [p. 21]

Can the value of a life be quantified? Despite the aversiveness of thinking about life in economic terms, the fact is inescapable that by our actions we put a finite value on our lives. Decisions to install safety features, to buy life insurance, and to accept a more hazardous job for high salary all carry implicit values for the worth of life.

Economists have long debated the question of how to quantify the value of a life or the value of a specified change in survival probability. Bergstrom (1974) argued that the best way to answer these questions is by observing the actual behavior of people trading risks for economic benefits. In this tradition, Thaler and Rosen (1975) studied salary as a function of occupational risk and concluded that a premium of about $200 (in 1967 dollars) per year was required to induce people in risky occupations to accept an annual increment in probability of accidental death of .001. From this, it can be argued that society should be willing to pay about $200,000 to avoid a statistical death. An extensive replication of the Thaler and Rosen study by Rappoport (1977) obtained a value of about $2,000,000. Howard (1980) inferred a value similar to Rappoport's when he asked people directly how much they would have to be paid to incur an additional .001 probability of death.

Should all modes of death be valued equally? Some proponents of nuclear power have complained about the public's apparent willingness to spend many times more money to prevent a fatality in the nuclear industry than to prevent other types of death. Is this so, and if so, is it reasonable?

Unfortunately, we do not have good answers to these questions. However, some data from the previously mentioned study by Fischhoff et al. (1978) are suggestive. In that study, members of the Eugene, Oregon, League of Women Voters indicated that an activity's acceptable risk level depended upon a variety of qualitative characteristics (such as voluntariness, dread, immediacy of consequences, familiarity, and controllability). It might be desirable to weight the aversiveness of deaths according to these characteristics. One could argue, for example, that deaths from risks imposed involuntarily should be counted more in the calculations of loss than deaths from voluntary risks. The size of the weighting factor would be a matter for society to decide.

Potential for catastrophe (the loss of a large number of lives at once) is another characteristic that might deserve higher negative weight, because of the horrific nature of catastrophes and because they may pose a greater threat to the survival of a community or society than do scattered individual deaths. Wilson (1975) has argued that the cost of n lives lost at once should be weighted by n^2 to take this into account, but Slovic, Fischhoff, and Lichtenstein (1980a, 1980b) have argued that a more complex model is necessary.

Another consideration in nuclear waste decisions is the treatment of delayed deaths. Should the death of an individual who succumbs to leukemia thirty years after exposure to radiation be assigned the same social cost as a death within weeks after exposure? Traditional economic theory argues that it is better to delay costs and, therefore, delayed deaths should be discounted or considered to be less serious. Any such discounting favors options that save bad consequences for later generations. Assuming a 5 percent annual discount rate, one immediate death would be equivalent to 1730 deaths two hundred years from now. Use of a discount rate has been vigorously protested (Lovins, 1976) as encouraging shoddy standards (for example, in bridges that collapse in twenty years, since a loss then wouldn't be worth much). Critics argue that the discount rate should be zero if we wish to minimize future generations' regret about our generation's choices (Schultze, 1974).

At present, there is no generally accepted method for weighing intergenerational benefits and costs. It has been suggested (National Academy of Sciences, 1975) that, until a method is developed, benefits and costs should be computed over a wide range of discount rates to see if the effects are significant.

Are there higher-order consequences that need consideration? The model illustrated in figure 5.5 may be inadequate because it considers only direct costs, such as harm to people and damage to the environment. What happens as a consequence of these first-order consequences may be even more important. For example, a reactor accident may also lead to the shutdown of the whole nuclear industry for some period of time. This second-order consequence would affect virtually every facet of our society (through even higher-order consequences such as power shortages, lost jobs, and unheated homes). Because society is so vulnerable to such events, even a small accident could result in massive social disruption. Although there is no inherent reason why higher-order effects could not be incorporated in an analysis, doing so is difficult and may introduce an even higher level of uncertainty. Nonetheless, it is unlikely that traditional (intuitive and political) decision procedures, unaided by decision analysis, would do any better job of considering these subtle factors.

FUTURE DIRECTIONS

For the immediate future, the primary tools brought to bear on risk management decisions will be various formal and semiformal analytic techniques. The results of these analyses will go through the political wringer, and some decisions will emerge. The research discussed in this chapter suggests a number of actions that interested citizens might take in order to involve themselves effectively in the decision-making process.

One action involves preparing themselves for the process. Preparation is ordinarily seen as a matter of boning up on the facts of an issue. The research on risk perception indicates a need to educate one's intuitions in order to understand these facts. That means learning to be wary of systematic biases, both those that come naturally and those that might be exploited by risk presenters. It means realizing where one's intuitions are not to be trusted and decision aids are needed. It means appreciating the limits of one's knowledge. H. G. Wells once observed, "Statistical reasoning will one day be as important for good citizenship as the ability to read and write." Perhaps that day has come.

The second set of tasks involves making formal analyses as useful as possible to the decision process. That means working to keep them honest. Just like eye-witness testimony, the evidence emerging from such analyses can be biased, incomplete, and half-truthful. Although such analyses may be the exception, a watchful eye is prudent. The usefulness of an analysis may also depend on its modesty. The recognition of uncertainty is a central facet of decision analysis. Nonetheless, analysts enmeshed in their work may not realize the limits of their own critical powers or the limits of the scientific knowledge on which they rely. Inadequate caution is encouraged by the pressures of public situations, with their demand for definitive statements, and the desire to affect decisions.

Just as performing an analysis requires skills from many disciplines, so is it difficult for any one individual to criticize all facets. Concerned citizens might formulate their own independent review teams with experts from different fields. This strategy has apparently been successfully implemented in Massachusetts (Smardon and Woodland, 1976–77).

Another task is to make sure that analysts make their assumptions explicit. There is no reason to expect most citizens to realize the ethical implications underlying the use of a positive social discount rate or to recognize which alternative courses of action have been summarily ignored in the analysis. The public should insist on being told what assumptions about life and data it is accepting along with the "facts" of an analysis.

We believe that the basic function of formal analyses should be to help policy makers gain insight into complex issues. In our view, the important element is not the bottom line, but the process of reaching it. Policy makers need to understand the assumptions and calculations that led to the result. The analysis should shift debate from the decision itself to the critical impacts and assumptions, highlighting the most sensitive issues for scrutiny.

Lovins (1976) warns about "the delicacy of the balance between drawing on expertise and smothering democracy with it" (p. 114). Accordingly, the public must work to keep societal decision processes open and responsive. Analysts and regulators are paid out of public funds. They should make their analyses comprehensible, solicit public input, and reflect public desires in their conclusions. A strong case can be made that the future of the democratic process is being shaped in these risk-management decisions. Excluding the public here may create a new technocratic elite, in effect declaring the public to be technically incompetent and disenfranchising it from a broad range of important decisions.

REFERENCES

Barrager, S. M.; Judd, B. R.; and North, D. W. 1976. *Decision Analysis of Energy Alternatives: A Comprehensive Framework for Decision Making.* Palo Alto, CA: Stanford Research Institute.

Bergstrom, T. C. 1974. Preference and choice in matters of life and death. Appendix 1 in Report No. ENG 7478, School of Engineering and Applied Science, UCLA, November 1974.

Blake, V. E. 1968. A Prediction of the Hazards from the Random Impact of Meteorites on the Earth's Surface. In *Aerospace Nuclear Safety Report SC-RR-68-838*. Albuquerque, N.M.: Sandia Labs.

Boffey, P. M. 1975. Nuclear War: Federation Disputes Academy on How Bad Effects Would Be. *Science* 190:248–50.

Borch, K. 1968. *The Economics of Uncertainty.* Princeton, NJ: Princeton University Press.

Bryan, W. B. 1974. Testimony before the Subcommittee on State Energy Policy, Committee on Planning, Land Use, and Energy, California State Assembly, February 1, 1974.

Coates, J. F. 1976. The Role of Formal Models in Technology Assessment. *Technological Forecasting and Social Change,* vol. 8.

Cohen, B. L. 1974. Perspectives on the nuclear debate. *Bulletin of the Atomic Scientists,* vol. 30, no. 9, pp. 35–39.

Cohen, J. J. 1972. A Case for Benefit-Risk Analysis. *Risk vs. Benefit: Solution or Dream,* ed. H. J. Otway. Report LA-4860-MS, Los Alamos Scientific Laboratory (available from the National Technical Information Service).

Comey, D. D. 1975. How We Almost Lost Alabama. *Chicago Tribune,* August 31, 1975, p. 2.

David, E. E. 1975. One-Armed Scientists? *Science* 189:891.

Fischhoff, B. 1975a. Hindsight≠Foresight: The Effect of Outcome Knowledge on Judgment under Uncertainty. *Journal of Experimental Psychology: Human Perception and Performance* 1:289–99.

______. 1975b. Hindsight: Thinking Backward? *Psychology Today* 8:70–76.

______. 1977a. Perceived Informativeness of Facts. *Journal of Experimental Psychology: Human Perception and Performance* 3:349–58.

______. 1977b. Cost-Benefit Analysis and the Art of Motorcycle Maintenance. *Policy Sciences* 8:177–202.

Fischhoff, B., and Beyth, R. 1975. "I Knew It Would Happen": Remembered Probabilities of Once-Future Things. *Organizational Behavior and Human Performance* 13:1–16

Fischhoff, B.; Slovic, P.; and Lichtenstein, S. 1977. Knowing with Certainty: The Appropriateness of Extreme Confidence. *Journal of Experimental Psychology: Human Perception and Performance* 3:552–64.

______. 1978. Fault Trees: Sensitivity of Estimated Failure Probabilities to Problem Representation. *Journal of Experimental Psychology: Human Perception and Performance* 4:342–55.

Fischhoff, B.; Slovic, P.; Lichtenstein, S.; Read, S.; and Combs, B. 1978. How Safe is Safe Enough? A Psychometric Study of Attitudes Towards Technological Risks and Benefits. *Policy Sciences* 8:127–52.

Gray, E. F. 1976. Remarks on the Problem of Nuclear Waste Management. *Proceedings of the 1976 Conference on Public Policy Issues in Nuclear Waste Management.* National Science Foundation, pp. 195–202. Washington, D.C.

Green, A. E., and Bourne, A. J. 1972. *Reliability Technology*. New York: Wiley.

Hall, W. K. 1975. Why Risk Analysis Isn't Working. *Long Range Planning* (December):23–29.

Herrero, S. 1970. Human Injury Inflicted by Grizzly Bears. *Science* 170:593–97.

Hohenemser, K. H. 1975. The Failsafe Risk. *Environment* 17:6–10.

Holdren, J. P. 1976. The Nuclear Controversy and the Limitations of Decision Making by Experts. *Bulletin of the Atomic Scientists*, vol. 32, no. 3, pp. 20–22.

Howard, R. A. 1968. The Foundations of Decision Analysis. *IEEE Transactions on Systems, Science and Cybernetics* 4:211–19.

______. 1975. Social Decision Analysis. *Proceedings of the IEEE* 63: 359–71.

______. 1980. On Making Life or Death Decisions. In *Societal Risk Assessment: How Safe Is Safe Enough?* Ed. R. C. Schwing and W. A. Albers, Jr.

Howard, R. A.; Matheson, J. E.; and North, D. W. 1972. The Decision to Seed Hurricanes. *Science* 176:1191–202.

Hynes, M., and Vanmarcke, E. 1976. Reliability of Embankment Performance Predictions. *Proceedings of the ASCE Engineering Mechanics Division Specialty Conference.* Waterloo, Ontario, Canada: University of Waterloo Press.

Kahneman, D.; Slovic, P.; and Tversky, A. 1982. *Judgment under Uncertainty: Heuristics and Biases.* New York: Cambridge University Press.

Kates, R. W. 1962. Hazard and Choice Perception in Flood Plain Management. Research Paper 78, Department of Geography, University of Chicago.

———. 1978. *Risk Assessment of Environmental Hazard.* New York: Wiley.

Katz, D. L., and West, H. H. 1975. The Overall-Problem—Risk/Benefit for LNG Shipping and Storage. In *Risk-Benefit Methodology and Application: Some Papers Presented at the Engineering Foundation Workshop, Asilomar, CA,* ed. D. Okrent. Report ENG-7598, School of Engineering and Applied Science, UCLA, December 1975, pp. 1–40.

Keeney, R. L., and Raiffa, H. 1976. *Decisions with Multiple Objectives: Preferences and Value Trade-offs.* New York: Wiley.

Lichtenstein, S., and Slovic, P. 1973. Response-Induced Reversals of Probabilities: State of the Art to 1980. In *Judgment under Uncertainty: Heuristics and Biases,* eds. D. Kahneman, P. Slovic, and A. Tversky. New York: Cambridge University Press.

Lichtenstein, S., and Slovic, P. 1973. Response-Induced Reversals of Preference in Gambling: An Extended Replication in Las Vegas. *Journal of Experimental Psychology* 101:16–20.

Lichtenstein, S.; Slovic, P.; Fischhoff, B.; Layman, M.; and Combs, B. 1978. Judged Frequency of Lethal Events. *Journal of Experimental Psychology: Human Learning and Memory* 4:551–78.

Lifton, R. J. 1976. Nuclear Energy and the Wisdom of the Body. *Bulletin of the Atomic Scientists,* vol. 32, no. 7, pp. 16–20.

Lovins, A. B. 1976. Comments in S. M. Barrager, B. R. Judd, and D. W. North, *Decision Analysis of Energy Alternatives.* Palo Alto, CA: Stanford Research Institute.

Lowrance, W. W. 1976. *Of Acceptable Risk.* Los Altos, CA: Wm. Kaufmann.

McGrath, P. E. 1974. Radioactive Waste Management: Potentials and Hazards from a Risk Point of View. Report EURFNR-1204 (KFK 1992), U.S.-EURATOM Fast Reactor Exchange Program, Karlsruhe, Germany.

Matheson, J. E., and Roths, W. J. 1967. Decision Analysis of Space Projects: Voyager Mars. Paper presented at the National Symposium "Saturn, Apollo and Beyond," American Astronautical Society, June 1967.

Matheson, J. E.; Cazalet, E. G.; North, D. W.; Rousseau, W. F.; and Howard, R. A. 1968. Decision Analysis of Nuclear Plants in Electrical System Expansion. Final Report, SRI Project 6469, Stanford Research Institute, Menlo Park, CA.

National Academy of Sciences. 1975. *Decision Making for Regulating Chemicals in the Environment.* Appendix H, pp. 163–96.

Nuclear Regulatory Commission. 1975. *Reactor Safety Study: An Assessment of Accident Risks in U.S. Commercial Nuclear Power Plants.* Washington, D.C. WASH 1400, NUREG-75/014.

______. 1978. *Risk Assessment Review Group Report to the U.S. Nuclear Regulatory Commission.* NUREG/CR-0400.

Otway, H. J., and Cohen, J. J. 1975. Revealed Preferences: Comments on the Starr Benefit-Risk Relationships. Research Memorandum 75-5, International Institute for Applied Systems Analysis, Laxenburg, Austria.

Otway, H. J., and Pahner, P. D. 1976. Risk Assessment. *Futures* 8: 122–34.

Pahner, P. D. 1975. The Psychological Displacement of Anxiety: An Application to Nuclear Energy. In *Risk-Benefit Methodology and Application: Some Papers Presented at the Engineering Foundation Workshop, Asilomar, CA,* ed. D. Okrent. Report ENG-7598, School of Engineering and Applied Science, UCLA, December 1975, pp. 557–80.

Pauker, S. G. 1976. Coronary Artery Surgery: The Use of Decision Analysis. *Annals of Internal Medicine* 85:8–18.

Primack, J. 1975. Nuclear Reactor Safety: An Introduction to the Issues. *Bulletin of the Atomic Scientists,* vol. 31, no. 9, pp. 15–17.

Rabinowitch, E. 1972. Living Dangerously in the Age of Science. *Bulletin of the Atomic Scientists,* vol. 28, no. 1, pp. 5–8.

Raiffa, H. 1968. *Decision Analysis.* Reading, MA: Addison-Wesley.

Rappoport, E. 1977. Ph.D. dissertation, Department of Economics, UCLA.

Ross, L. 1977. The Intuitive Psychologist and His Shortcomings: Distortions in the Attribution Process. In *Advances in Experimental Social Psychology,* ed. L. Berkowitz, New York: Academic Press, pp. 173–220.

Rowe, W. D. 1977. *An Anatomy of Risk.* New York: Wiley.

Schultze, W. 1974. Social Welfare Functions for the Future. *American Economist* 18:70–81.

Schwing, R. C., and Albers, W. A., eds. 1980. *Societal Risk Assessment: How Safe is Safe Enough?* New York: Plenum.

Sinsheimer, R. F. 1971. The Brain of Pooh: An Essay on the Limits of Mind. *American Scientist* 59:20–28.

Slovic, P., and Fischhoff, B. 1977. On the Psychology of Experimental Surprises. *Journal of Experimental Psychology: Human Perception and Performance* 3:544–51.

Slovic, P.; Fischhoff, B.; and Lichtenstein, S. 1977. Behavioral Decision Theory. *Annual Review of Psychology* 28:1–39.

——1980a. Facts and Fears. Understanding Perceived Risk. In *Societal Risk Assessment: How Safe Is Safe Enough?* Ed. R. C. Schwing and W. A. Albers, Jr.

——1980b. Perceived Risk and Quantitative Safety Goals for Nuclear Power. *Transactions of the American Nuclear Society* 35:400–01.

Slovic, P.; Kunreuther, H.; and White, G. F. 1974. Decision Processes, Rationality, and Adjustment to Natural Hazards. In *Natural Hazards, Local, National and Global,* ed. G. F. White. New York: Oxford University Press.

Smardon, R. C., and Woodland, R. B. 1976–77. Some Preliminary Results of an Environmental Impact Report Review Process. *Journal of Environmental Systems* 6:209–28.

Starr, C. 1969. Social Benefit Versus Technological Risk. *Science* 165:1232–38.

Thaler, R., and Rosen, S. 1975. The Value of Saving a Life: Evidence from the Labor Market. In *Household Production and Consumption,* ed. N. E. Terleckyj, National Bureau of Economic Research.

Tversky, A., and Kahneman, D. 1974. Judgment under Uncertainty: Heuristics and Biases. *Science* 185:1124–31.

U.S. Government. 1976. *Teton Dam Disaster.* Committee on Government Operations, Washington, D.C.

Weinberg, A. M. 1976. The Maturity and Future of Nuclear Energy. *American Scientist* 64:16–21.

Wilson, R. 1975. Examples in Risk-Benefit Analysis. Conference on Advanced Energy Systems. *Chemtek* 5:604.

Zebroski, E. L. 1975. Attainment of Balance in Risk-Benefit Perceptions. In *Risk-Benefit Methodology and Application: Some Papers Presented at the Engineering Foundation Workshop, Asilomar, CA,* ed. D. Okrent. Report ENG-7598, School of Engineering and Applied Science, UCLA, pp. 633–44.

EDWARD J. WOODHOUSE

6 THE POLITICS OF NUCLEAR WASTE MANAGEMENT

This chapter analyzes the political process by which decisions will be made about the management of radioactive wastes. The first section of the chapter describes the federal agencies, interest groups, elected officials, and other political forces that are likely to influence upcoming decisions; and it discusses the role the public can expect to play in this process. The second section of the chapter examines a variety of institutional problems and other constraints that are likely to be obstacles in formulating and implementing a program for handling radioactive wastes. Lacking prescience, it is impossible to predict with confidence how satisfactory the U.S. waste management program will prove in the future. Even if all the technical problems can be solved, however, the analysis suggests political reasons to fear that radioactive wastes may remain too hot to handle.

WHO WILL MAKE THE DECISIONS?

In the fragmented political structure of the United States, major issues are seldom resolved by a single political institution or with a single yes-no decision. Instead, major decisions evolve gradually as partial, interim choices are made by many individuals and institutions in a long and complex decision process.

My thanks to numerous former colleagues at the Institution for Social and Policy Studies, Yale University, for insightful comments on an earlier draft of this chapter. Leroy C. Gould, Charles E. Lindblom, Vera McCluggage, and Joseph Morone were particularly helpful. Jeanne Hurlbert, College of William and Mary, provided excellent research assistance. The work was supported in part by a grant to ISPS from the U.S. Energy Research and Development Administration.

To complicate matters further, most of the political institutions responsible for nuclear waste have undergone a series of reorganizations; and there may be further changes. So the following is a sketch only of the main features of the institutions that will be most responsible for shaping U.S. policy on nuclear wastes in the near future. They include several agencies in the executive branch of the federal government, Congress, the courts, state and local governments, industry, scientific organizations, public interest groups, and the general public.

The Federal Agencies

From 1947 until 1974, operating under substantial secrecy, the Atomic Energy Commission (AEC) had virtually sole programmatic responsibility for U.S. nuclear energy activities. When the AEC was dissolved by Congress in 1974, its tasks were divided between two new organizations. The Nuclear Regulatory Commission assumed responsibility for regulating nuclear power; research, promotion, and military aspects of nuclear energy were taken over by the Energy Research and Development Administration (ERDA), subsequently merged into the Department of Energy.

The U.S. Department of Energy (DOE), created in 1977, has the major governmental responsibility for all forms of energy, including nuclear power and the waste that it generates. Congress has charged DOE with responsibility for developing a facility capable of receiving and storing high-level nuclear wastes. The Waste Management Division of DOE bears primary responsibility for locating suitable sites for a waste repository and for developing an environmentally safe waste management technology. Since DOE does not have sufficient expertise to carry out all the work required for this task, much of it is conducted at national laboratories such as Oak Ridge or contracted out to university researchers and private industry. The Battelle Institute, for example is DOE's prime contractor for the Nuclear Waste Terminal Storage program. Pending establishment of permanent waste repositories, DOE owns and operates interim burial grounds and storage facilities for radioactive wastes generated by military weapons development and other DOE research. The department is required by law to prepare and circulate environmental impact statements analyzing the probable effects of proposed alternatives for managing radioactive wastes and then to hold public hearings where organizations or individuals can make their views known.

The five-member Nuclear Regulatory Commission (NRC) and its staff regulate all civilian nuclear activities. A NRC license is required for any use of nuclear materials. Thus, the NRC oversees the mining of uranium, the production of fuel rods, the construction and operation of nuclear-fueled electric power generating plants, and the transportation and reprocessing or disposal of radioactive waste, as well as the operation of nuclear facilities used for medical therapy and research. The agency cooperates with the states in many of these activities.

In carrying out its regulatory function, the NRC is attempting to develop a comprehensive regulatory program applicable to the period of terminal storage that follows solidification or other processing of nuclear wastes. The commission develops procedures for assessing risks from geological isolation of high-level and transuranic wastes, prepares regulations on the acceptable form of solidified high-level waste and its packaging, and contracts for research on waste problems. While DOE will construct and operate waste repositories, the NRC must license them.

The Environmental Protection Agency (EPA) is responsible for the promulgation and enforcement of antipollution regulations, including those setting radiation protection standards. EPA decides what the numerical limits should be on the amount of radioactivity allowable from uranium and phosphate mining and milling wastes, low-level wastes, high-level wastes, wastes from decommissioning, and other aspects of the nuclear fuel cycle. EPA standards are not site or method specific; they are general environmental standards. It is then up to the NRC and the DOE to see that the standards set by EPA are met.

Other federal departments and agencies participate in radioactive waste policy in more limited ways. For example, the Department of Transportation is responsible for regulations governing the transport of wastes; the U.S. Geological Survey helps assess the suitability of potential repository sites; the Department of State is concerned with foreign policy implications, and the Department of Interior is involved when waste management impinges on Native American or federally owned lands.

Although interagency cooperation is not a hallmark of the federal government, there have been beginnings of such cooperation in the waste management field. The Office of Management and Budget in the Ford administration chaired an effort to coordinate all federal research, development, and evaluation projects on nuclear waste management. President Carter went even further in establishing an Interagency Review Group composed of representatives from fourteen departments and agencies; the IRG report exerted considerable influence in shaping the Carter administration's approach to the waste management program (IRG, 1979; Moss, 1982).

U.S. Congress

Prior to 1977, full congressional responsibility for nuclear energy matters rested with a prestigious House-Senate Joint Committee on Atomic Energy (JCAE). That committee was abolished by the 95th Congress, in part because of sentiment that the committee had failed to raise timely questions about the safety and economics of nuclear energy (Del Sesto, 1980). The joint committee's jurisdiction was divided among several committees in each house.

In the House of Representatives, the Interior Committee now handles domestic regulation of nuclear energy, including the annual budget authorization for the Nuclear Regulatory Commission. The Energy and Commerce Committee has jurisdiction over the regulation and oversight of nuclear facilities. A third House committee, Science and Technology, is responsible for nuclear research. Interior has tended to take positions preferred by environmental groups while Science has been more responsive to industry. On some issues, such as the proposed Waste Isolation Pilot Plant in New Mexico, all three committees claim jurisdiction and report conflicting bills to the full House.[1]

In the Senate, the Committee on Energy and Natural Resources has jurisdiction over energy development, fossil as well as nuclear, and has been the most influential in shaping Senate policy on radioactive waste management. Recent committee chairmen have been strong advocates of nuclear power. The Committee on Environment and Public Works has jurisdiction over nuclear safety and environmental protection; it usually has been headed by moderate environmentalists. The Senate Governmental Affairs Committee shares jurisdiction concerning nuclear waste storage.[2]

Recent efforts of these committees to formulate a comprehensive plan for radioactive waste management are discussed below.

State and Local Governments

Generally quiescent or supportive of federal efforts in the early years of nuclear power, numerous states and municipalities began to express concerns about nuclear safety in the late 1970s. For example, more than forty municipalities, including New York City, have passed laws restricting transportation of radioactive wastes within city limits; a number of states have

1. The differences among House committees are due in part to their chairmen. Environmentalist Morris K. Udall (D-Ariz.) has chaired the Interior Committee and its key Energy & Environment subcommittee since 1977, while Energy and Commerce is chaired by Harley O. Staggers (D-W. Va.), and Science and Technology is under Don Fuqua (D-Fla.). Other House committees that share jurisdiction on some nuclear issues include Armed Services, Government Operations (especially the subcommittee on Environment, Energy and Natural Resources, with oversight of DOE and NRC), and Appropriations.

2. Senate chairmanships have been as follows: Energy and Natural Resources: Henry M. Jackson (D-Wash.) through 1980; James A. McClure (R-Idaho) 1981–82; Environment and Public Works: Jennings Randolph (D-W. Va.) through 1980, Robert T. Stafford (R-Vt.) 1981–82. Other Senate committees sharing jurisdiction on some nuclear waste matters have included Commerce, Science and Technology, Transportation (waste transport), Foreign Relations (nuclear export), Armed Services, and Appropriations.

done likewise. In California, a 1976 law prohibited construction of additional nuclear power plants pending development of a federal program to manage nuclear wastes. A Vermont law passed in 1977 prohibited construction of waste storage facilities without approval from the state General Assembly. Officials, press, and local residents in Michigan raised sufficient political uproar that ERDA promised the governor a veto over establishment of any waste facility in that state. State geologists, the governor, and congressmen from Kansas induced AEC to abandon plans for a waste repository in 1971.

The extent to which state and local governments can make their actions stick will depend on court interpretations, executive policy, and congressional action. The NRC and Department of Transportation, for example, issued regulations in 1980–81 that prevent localities from blocking transportation of radioactive waste on approved routes (Green and Zell, 1982). However, the Carter administration promised to states and localities a right of "consultation and concurrence" on waste policy, although it was never clear what the terms meant (Varanini, 1982). The Reagan administration has backed away from even the vague Carter promise, but Congress is certain to be sensitive to political pressure from state officials and voters. Most versions of the waste management bills considered by Congress since the late 1970s have contained provisions under which a state could veto siting of a high-level waste repository under certain circumstances. But whether or not formal rights ultimately are accorded the states when a federal high-level waste plan is finally adopted, states and localities will retain at least some measure of informal political influence.

States have exercised even more influence over the management of low-level wastes. At one time there were six sites that accepted low-level commercial wastes. Two closed in the late 1970s for economic and technical reasons. Kentucky placed a 10 cents per pound tax on radioactive materials that led to the closing of the Maxey Flats burial site in 1977. Reacting to fires and other problems apparently caused by inadequate federal enforcement of packaging and transportation regulations, the governor of Nevada has closed the Beatty disposal site for various periods and has threatened to terminate it altogether. The governor of South Carolina has cut in half the amount of waste that the Barnwell site, the nation's largest, is allowed to accept. Finally, voters in Washington state attempted through a 1980 referendum to close the Hanford disposal site to out-of-state nuclear wastes of nonmedical origin.

Partly in response to such state actions, in early 1980 President Carter established a State Planning Council on Radioactive Waste Management. The council contributed to pressures that led Congress to pass the Low-Level Radioactive Waste Policy Act of December, 1980 (State Planning Council, 1981).[3] Henceforth, each state will be responsible for providing disposal capacity for low-level wastes generated within its borders, presumably in regional facilities rather than a separate facility in each state. Military and other federal low-level wastes are excluded from the arrangement.

Federal Courts

The major rulings to date by federal courts on nuclear issues have concerned nuclear power plants much more than radioactive waste. But the pattern of decisions and the legal rationale utilized by the courts suggests the way in which future challenges regarding waste disposal will be handled.

An important 1971 decision denied the Minnesota Pollution Control Agency the right to impose stringent standards regulating the level of radioactive liquid and gaseous discharges by a Northern States Power Company nuclear plant. As reiterated in a later case, in *Northern States* the court held that the Atomic Energy Act "created a pervasive regulatory scheme, vesting exclusive authority to regulate the discharge of radioactive effluents from nuclear power plants in the AEC, and pre-empting the States from regulating such discharges" (U.S. Supreme Court, 1976, p. 16). In 1976, the Supreme Court declared nuclear power plants exempt even from federal environmental legislation. Although the Federal Water Pollution Control Act

3. The Low-Level Radioactive Waste Policy Act of 1980 is Public Law 96–573 (S.2189), December 23, 1980, 94. Stat. 3347.

makes it unlawful to discharge pollutants (specifically including "radioactive materials") into navigable waters without a permit from the Environmental Protection Agency, the EPA was held to have no authority over radioactive emissions from nuclear plants (U.S. Supreme Court, 1976). Federal judges in 1979 and 1980 struck down all California laws regulating nuclear power plants on the grounds of federal preemption (U.S. District Court, 1979, 1980). The same rationale was applied to nuclear waste more specifically when a federal court overturned the 1980 Washington State limitation on out-of-state low-level wastes (U.S. District Court, 1981). In short, unless new federal laws are approved giving additional authority to states or localities, the federal courts will probably allow only the Nuclear Regulatory Commission to regulate nuclear power and its resulting wastes (Karpinski, 1976; Murphy and LaPierre, 1976; Green and Zell, 1982).

On the other hand, courts have upheld procedural challenges where utilities, the NRC, or the Department of Energy did not comply with the process by which nuclear facilities are supposed to be evaluated and approved. The net effect of court intervention, therefore, has been primarily to slow down the development of new sites, to induce those responsible to weigh environmental considerations a bit more heavily, and to complicate the bureaucratic procedures that have to be followed. Not surprisingly, this focus on procedure has led to dissatisfaction with the slowness of the licensing process and to proposals to speed it up. The most substantial of these proposals, President Carter's "Fast Track Energy Board," was defeated in Congress, but a minor speed-up bill passed easily in 1981 and will allow the NRC to grant interim operating licenses to new nuclear power plants before the full regulatory proceedings have been completed.[4]

4. Slow NRC licensing procedures probably have been a more significant factor than anything the courts have done, and NRC delays appear to be due to important substantive issues more than to legalistic strategies by nuclear opponents (Cohen, 1979).

Industry

Utility companies, reactor manufacturers, uranium suppliers, and a host of smaller firms (such as those that build lead casks for shipping radioactive wastes) have a financial stake of several hundred billion dollars in the future of nuclear power and nuclear waste. Many of these nuclear corporations act in politics on their own, through trade associations such as the Atomic Industrial Forum, and through industry-financed interest groups such as Americans for Energy Independence. These organizations take out ads in the media, issue press releases, hire lobbyists, help finance political campaigns, and otherwise attempt to influence public opinion and political decision making.

In the referendum campaigns on nuclear issues waged in various states, for example, the financial contributions of industry have been substantial. The Atomic Industrial Forum alone spent an estimated $1.5 million on such electoral conflicts in 1976, and in that year corporations such as Bethlehem Steel and Exxon Nuclear contributed an estimated total of $2 million for the battle over California's antinuclear Proposition 15 (Fallows, 1979, pp. 96, 109). Out-of-state nuclear utilities contributed much of the $800,000 spent to help defeat a 1980 Maine referendum that could have shut down that state's only nuclear reactor; in contrast, antinuclear forces raised only $175,000, mainly in small contributions (*New York Times*, 1980b). While antinuclear forces have been outspent by margins as high as 25 to 1 in some states, the nuclear industry argues that the purchase of expensive media advertising is necessary to counteract an antinuclear bias in news coverage and to compensate for the unpaid volunteers who marshall votes against nuclear power (*New York Times*, 1975, 1976).

Expertise, however, is probably as vital a resource as money in policy deliberations, and the people who know the most about nuclear waste are those employed by nuclear-related industries. Although nonpartisan expertise is available from some scientists at universities and private research organizations, industry can often bring concentrated expertise to bear in ways that give their cause a decided advantage. For example, at the 1977 licensing hearings on the Three Mile Island reactor that later became controversial, some fifty-five expert witnesses testified in support of the application while a coalition of environmental groups was forced to make its case with only one expert (House Committee on Interior and Insular Affairs, 1978, p. 924). This disproportionate access to funds and expertise is part of what has been called the privileged position of business in the political arena (Lindblom, 1977).

The Scientific Community

In many spheres of modern society, scientists and social scientists with specialized knowledge participate influentially in decision making. This influence is channelled in part through an elite scientific organization, the National Academy of Sciences-National Research Council. The NAS operates directly as an advisory arm of federal executive agencies which commission studies and indirectly through its influence on scientists in academia, government, and business. Early NAS reports, beginning in 1957, were relatively critical of the AEC's handling of radioactive wastes (Committee on Waste Disposal, 1957; Hubbert, 1962). The 1966 report was actually suppressed by the AEC, and the responsible NAS committee was disbanded. A new committee, formed in 1968 and composed of a high percentage of scientists with close ties to the nuclear industry, for a time issued reports more supportive of the AEC (Boffey, 1975). Recent NAS panels have had relatively diverse membership, and their conclusions have been challenged by both environmentalists and those responsible for nuclear waste programs (*New York Times,* 1979).

Partly to challenge the viewpoint offered by NAS and industry scientists, the Union of Concerned Scientists was formed to conduct independent studies on nuclear power and other technical subjects. In addition to publishing reports generally critical of the nuclear waste program, UCS has testified at public hearings and undertaken other activities designed to improve the scientific expertise available to the environmental cause. With over 75,000 members and a substantial budget, the organization's studies have been well enough done to be credible to outside scientists and to achieve attention by the press and by decision makers.

Other vehicles for transmission of scientific expertise into the political process are as various as the Office of Technology Assessment, the General Accounting Office, congressional agencies that provide analysis and recommendations to Congress, presidential science and technology advisors, and the Scientists Institute for Public Information, which provides information primarily for the press.

The Public

In addition to public opinion polls and letters to public officials or the media, there are several avenues by which members of the public might participate in decision making about radioactive waste disposal.

Because few candidates are likely to make a major issue of the subject, elections for public office are not a promising mechanism for participation.[5] Nuclear power and nuclear waste, however, have been issues in numerous state-level initiatives and referenda. Citizens in seven states filed petitions with sufficient signatures to require referenda in 1976; none were successful. The issue did not appear on state-wide ballots during the next two years, but residents of thirty-five Vermont communities adopted local ordinances banning nuclear reactors and nuclear waste. Montana voters in 1979 adopted a ban on nuclear power plant construction until adequate means of waste disposal are developed. Of six state ballot measures in 1980, the Washington state referendum passed, while voters in Maine, Montana, Missouri, Oregon, and South Dakota rejected proposed restrictions on nuclear power, uranium mining, and

5. Under the right conditions, of course, anything can become a major electoral issue. In the 1976 Swedish election, an opposition party that opposed nuclear power ousted the Social Democrats who had governed for more than thirty years.

wastes. About half the states have no provisions for such referenda at present; neither does the federal government.

Public interest groups offer an alternative route for attempting to influence national policy on the subject. Although they lack formal authority, nonprofit public interest groups have already been influential in nuclear policy making by providing members of Congress with sophisticated evidence and argument, by scrutinizing the work of executive agencies, and by initiating litigation in federal courts. The efforts of the Union of Concerned Scientists have already been mentioned. Other such groups include Common Cause, the Natural Resources Defense Council, the Sierra Club, and Friends of the Earth; all of these tend to be critical of DOE's nuclear programs. Perhaps because pronuclear interests have been well represented by industry, Congress, and federal nuclear agencies, there are fewer groups organized to support expansion of nuclear energy or support of current waste management programs; one that does is Americans for Energy Independence.

Although relatively infrequent recently in the U.S., sit-ins and other protest activities have a long history in American politics, dating from the Boston Tea Party, and have also been common in Europe. Some 30,000 demonstrators clashed with French police in 1977 near Malville, site of the "Super Phenix," France's giant plutonium breeder reactor. More than 50,000 gathered in Hanover, West Germany, in 1979 to protest construction of a $7 billion nuclear waste reprocessing plant and waste repository; and over 100,000 marched in an antinuclear demonstration that year in Bonn (Nelkin and Pollak, 1981; Zinberg, 1982). In the U.S., approximately 1,500 were arrested during 1977 demonstrations against the Seabrook nuclear reactor, an estimated 25,000 attended a 1980 rally at the Washington Monument, and some 2,000 were arrested in 1981 while protesting the scheduled startup of the flawed Diablo Canyon power plant in California. But all of these protests against civilian nuclear power have now been overshadowed by the giant 1982 demonstrations in both Europe and the U.S. concerning nuclear weaponry.

Finally, increased political conflict over nuclear and other environmental issues, together with marked decline in public trust in government, has led both the U.S. and western European governments to increase opportunities for public review of governmental plans (Deese, 1982; Zinberg, 1982). Congress has written such requirements into recent laws, and several presidents have added executive orders on the subject. President Carter's proposal for a comprehensive waste management program decreed, for example, that

> it is essential that all aspects of the waste management program be conducted with the fullest possible disclosure to and participation by the public and the technical community. I am directing the departments and agencies to develop and improve mechanisms to ensure such participation and public involvement consistent with the need to protect national security information. [Carter, 1980, p. 414]

In principle, there are numerous mechanisms by which public participation could be sought. In practice, the dominant mechanism used by DOE, NRC, EPA, and other federal agencies is the public hearing. Rule-making hearings typically are held in Washington, D.C., conducted in quasi-legal format by a hearing examiner, with participation by legal counsel and professional representatives of industry and other interests. A specific set of standards or regulations is at issue. "Public forums," "public workshops," and other less formal types of public hearings tend to be held away from Washington, D.C., attract a somewhat less expert audience, and allow discussion of broader questions in an exploratory fashion.[6]

What effects will these forms of public participation have on the upcoming decisions about the U.S. radioactive waste program? The general public is likely to have substantial influence on certain aspects of waste disposal and virtually no influence over others. For example, the public surely will have almost no influence in choosing among the specific technical alternatives for containing radioactivity. These decisions will be dominated by the weight of engineering and scientific opinion at the time the choices have to be made, by staff decisions in the Depart-

6. The usefulness of public hearings is discussed later in this chapter.

ment of Energy and Nuclear Regulatory Commission, and by the political judgment of a handful of top political leaders.

In choosing the location of nuclear waste repositories, on the other hand, local public sentiment could be critical. Citizens in areas of proposed repositories may well come to feel so threatened that nuclear issues will become a highly salient political dispute. There are enough potential waste repository sites, however, for the Department of Energy to pull out of areas where there is the strongest local protest. Only if there are sizeable pockets of local protest at all the geologically suitable repository sites, consequently, would the overall direction of national nuclear waste policy necessarily be influenced by local political activity.[7]

On a third nuclear waste issue the general public also has the potential for considerable influence: the question of how much more waste will be produced. Although direct responsibility rests with electrical utility companies, who decide how many new nuclear-powered generating plants to build, their decisions are influenced by the policies of the president and DOE secretary as well as by obstacles to licensing posed by the NRC and environmental groups. Indirectly, consumers and industries that use electricity exert substantial influence on utility decisions according to how much electricity they demand. If consumers use increasing amounts of electricity, they may thereby be "voting" indirectly for increased nuclear power and, hence, increased radioactive waste.

CONSTRAINTS TO INTELLIGENT DECISION MAKING

Until the mid-1970s, policy making on nuclear power and nuclear waste was dominated by a "subgovernment" composed of the Joint Committee on Atomic Energy, the Atomic Energy Commission, and the nuclear industry. As described by one political scientist,

7. Of course, any local protest that contributes to further delay in repository siting and construction may indirectly cast additional doubt on the future of civilian nuclear power.

complacency and lack of scrutiny by Congress, several presidents, the press, and the general public allowed nuclear . . . policy to be made in secret by a relatively small number of individuals and groups which together constituted a cozy subgovernment interested chiefly in promotion rather than regulation. [Temples, 1980, p. 245]

The resulting radioactive waste management program is now widely considered to have been seriously deficient. President Carter acknowledged that "past governmental efforts to manage radioactive wastes have not been technically adequate. Moreover, they have failed to involve successfully the States, local governments, and the public in policy or program decisions" (Carter, 1980, p. 412).

Among other problems with the program were the following:

1. The priority given to waste management was very low, as reflected in near-trivial budget outlays of less than .1 percent of the AEC budget prior to the early 1970s (*Wall Street Journal,* 1971; Metlay, 1978; IRG, 1979, p. 4).

2. Scientific research was confined much too narrowly; as the IRG Report stated, there were "inadequate perceptions of the [need for] . . . unprecedented extension of capabilities in rock mechanics, geochemistry, hydrogeology, and long-term predictions of seismicity, volcanism, and climate" (IRG, 1979, pp. 2–3).

3. The AEC generally discouraged exploration of alternatives to established nuclear policy; dissenting AEC scientists found their research support reduced and their projects reassigned (Metzger, 1972, pp. 256–58; Lewis, 1972).

4. Safety and environmental issues often were treated as secondary issues; for example, although types of reactors differ substantially in safety and in the quantities and types of waste products they generate, a 1964 choice among five potential breeder reactor technologies was made on the basis of "the state of [each technology's current] development, utility interest, . . . and the project costs to the Government" (Joint Committee on Atomic Energy, 1966).

5. Safety and environmental research appears to have been used by AEC administrators more to justify decisions already made than to guide the agency's deliberations. For example, the breeder reactor became a top priority about 1964, then a cost-benefit analysis attempted to justify the expense (1969), and finally a study of safety and environmental factors attempted to show that the breeder would be safe (1975).

6. Federal nuclear studies have often been of poor quality, and the errors have tended to slant their conclusions in support of already-established policies. A 1975 fission safety report, intended to be the definitive study on the subject, was widely criticized in the scientific community for using an outdated and unreliable method for estimating the failure rate of reactor components and for numerous other deficiencies (Nuclear Regulatory Commission, 1975; Von Hippel, 1977; Study Group on Light Water Reactor Safety, 1975). Deficiencies in breeder reactor reports had the effect of inflating projected benefits or reducing projected costs (*Science,* 1974; GAO, 1975).

7. Federal agencies have gone to great lengths to deny the uncertainty inherent in nuclear technology. Comic books and glossy publications have extolled the virtues of nuclear power, and ERDA went so far as to distribute 78,000 copies of a promotional pamphlet apparently attempting to turn voters against a 1976 California referendum proposal on nuclear power. According to the General Accounting Office, a nonpartisan Congressional agency, the ERDA pamphlet was inaccurate "propaganda" that "misleads the reader into believing that technical, environmental, and social problems of storing high and low level wastes have been solved" (GAO, 1976, p. 10).

Numerous other shortcomings could be enumerated from the political history of U.S. radioactive waste management efforts, but the basic point is obvious: some forty years after the first production of radioactive waste by the Manhattan Project, the U.S. still lacks viable long-term capacity for isolating wastes from the biosphere.[8]

The current political situation is very different from that of the "cozy subgovernment" that prevailed in the 1950s and 1960s. No less than fourteen agencies shared in preparation of the IRG report that shaped Carter administration policy on radioactive waste. Some 3300 individuals in government, industry, environmental groups, and elsewhere offered comments on the draft report before it was made final. No fewer than six congressional committees exercise jurisdiction over nuclear

8. There is no book-length treatment of the political history of radioactive waste management, but short, selective histories with ample material include Metlay (1978), Fallows (1979), and Greenwood (1982). A somewhat more partisan view is Lipschutz, 1980, especially chapter 4. Many General Accounting Office reports offer useful historical information, including GAO 1968, 1974, and 1979.

waste management. Presidents are more deeply involved. Governors and other state and local officials are concerned and active. The press is vigilant, perhaps overly so on this issue compared with other social problems. The nuclear industry and its consultants no longer have a complete monopoly on expertise about the subject. In short, the changes have been profound, and nuclear critics err if they assume that the future will simply repeat the past.

Nevertheless, some constraints and obstacles do remain in the process by which waste management decisions will be made and implemented; in fact, certain of the changes have led to new problems.

Congressional Committees

During the Joint Committee's heyday, virtually everything it recommended passed Congress quickly. With numerous committees now sharing responsibility, conflict among them is inevitable. For example, because the Waste Isolation Pilot Plant proposed for Carlsbad, New Mexico, would store military wastes, the House Armed Services Committee has shared jurisdiction. The committee, however, has refused to fund the project if WIPP is required to obtain a license from the NRC; committee members are concerned that the necessary hearings might set a precedent that could open the way for future citizen challenges of military projects. The House Interior and Science committees, on the other hand, refuse to fund the project unless it is licensed: they consider the NRC licensing process an indispensable safeguard. Hence, there has been a political impasse.

More generally, complex jurisdictional and substantive disputes among committees and between House and Senate have contributed to the long delay in passing an overall federal plan for radioactive waste management. After years of preliminary skirmishing, the Senate passed one version of a Carter administration plan for waste disposal in July, 1980, with a delicate compromise on the military waste issue. The House took until early December to work out a plan acceptable to its various committees. Then, while the House and Senate versions were in conference committee, the Senate compromise came apart and the bill, except for a minor provision dealing with low-level wastes, was killed.

The scenario was similar in the 97th Congress. The House Interior and Science committees each reported a bill relatively early in the session, and the Senate passed a compromise bill in April, 1982. But the House Energy and Commerce committee delayed action in spite of pressure from industry, administration, and congressional sources. As of this writing, therefore, final passage of a comprehensive radioactive waste plan still appears unlikely. The seemingly endless delay and eventual compromises that result from fragmentation of congressional authority may be helpful or at least harmless in some instances; but there are almost certain to be times when such fragmentation results in less satisfactory public policies.

The increasing fragmentation of committee jurisdictions also may reduce the adequacy of congressional supervision of the radioactive waste program. "However irresponsible and biased the JCAE's performance might have been," Temples argues, "it at least offered a vehicle for centralizing congressional monitoring and oversight of the NRC in one place" (1980, p. 259). With so many committees and subcommittees, none can make credible threats or promises; hence each committee's bargaining power with the responsible federal agencies is reduced. Numerous committees carried on simultaneous investigations on the Three Mile Island incident, for example, but the useful information that resulted was not translated by Congress into clear directives for the NRC. Ironically, then, it can be argued that "Congress's very success in challenging the nuclear subgovernment by phasing out the JCAE may have done more in the long run to hamper strong, coordinated, effective legislative oversight of nuclear energy policy than to improve it" (Temples, 1980, p. 259).

Federal Agencies

Since about 1970, waste disposal has become a higher priority of the federal nuclear agencies: budgets have increased steadily, research is becoming more diverse, and planning is proceeding. However, some problems remain. After several years of planning and research, the AEC had to be informed in 1971 by the Kansas Geological Survey, a local salt company, and an outside consultant that a Lyons, Kansas, site chosen for a permanent waste repository was unsuitable because of the possibility of subsidence from salt mining, underground water, and twenty-nine gas and oil boreholes in the area (Metlay, 1978, pp. 5–7).

Following the Lyons embarrassment, planning began for a Retrievable Surface Storage Facility to contain high-level wastes for up to one hundred years, but a very negative environmental impact statement by EPA ended that plan. The primary basis for the low EPA rating was apparently the agency's judgment that such an "interim" facility might become permanent because of the financial and other difficulties of moving such a large accumulation of wastes. That sort of political judgment is usually seen as more appropriate for Congress and the president than for an EPA environmental impact assessment. Nonetheless, the project was thereby terminated. After that, the concept of away-from-reactor, interim surface storage was proposed, but the federal agencies have been unable so far to convince Congress of the worth of the plan.

Meanwhile, scientific opinion has shifted away from salt as a disposal medium, and the 1979 IRG report recommended an approach that would allow research to move ahead slowly on multiple disposal media and multiple potential sites. President Carter adopted this approach and called it "the nation's first comprehensive plan for disposing of high-level wastes" (Carter, 1980, p. 412). Although by no means an absurd decision, an uncharitable view might hold that the "plan" is little more than a nicely phrased way to delay taking tangible action on radioactive waste until further research clarifies the options.

During this same period, DOE has been developing plans for the Waste Isolation Pilot Plant discussed above. But in addition to congressional opposition, it turns out that there may be geological problems: several groundwater sources, faults penetrating the Carlsbad site, low-level seismic activity in the surrounding region, and potential for mineral exploitation that would violate NRC guidelines against locating a repository at any site that might be "an attractive target for future generations seeking natural resources" (1977, p. 5). President Carter decided to cancel the WIPP project, but work on it has continued under the Reagan administration.

There is recent evidence also that DOE's Office of Environment continues to play the relatively small role in decision making that characterized environmental offices in ERDA and in the AEC (GAO, 1980b).

Nor has the NRC won high marks of late. In 1978, the NRC's Advisory Committee on Reactor Safeguards reported that the agency's research program on radioactive wastes was "uncoordinated and unfocused." NRC's radioactive waste programs, moreover, continue to receive lower budget priorities than do many of the agency's other functions. In 1980, waste activities in the NRC's Office of Nuclear Material Safety and Safeguards ranked sixteenth in agency-wide priority, while waste management in the Office of Nuclear Regulatory Research ranked thirty-seventh (GAO, 1980a). Although waste management budgets have been increasing and there have been recent improvements in NRC performance, a 1980 GAO report found that "the Commission's regulatory performance can be characterized best as slow, indecisive, cautious—in a word, complacent" (GAO, 1980a, p. ii).

The President's Commission that investigated the Three Mile Island accident concluded that the NRC lacked the necessary organizational and management capabilities to pursue safety goals effectively (President's Commission, 1979). Abolition and replacement of the NRC in its present form was recommended. Although most NRC commissioners, key congressmen, and the President opposed such major restructuring, the Three Mile Island accident did lead the NRC to make some meaningful changes. Nevertheless, more than eighteen months after the accident, President Carter's newly created Nuclear Oversight Safety Committee (with a majority criticized as pronuclear by environmentalists) released a letter to President Carter that condemned what the committee termed a "business as usual mind-set" at the NRC. Specific criticisms included a list of unresolved safety issues that the committee said had lingered in regulatory limbo for 5–10 years (*New York Times*, 1980c).

Thus, while much has changed for the better, many thoughtful observers of the federal nuclear agencies remain highly critical of their performance.

Constraints of Budget and Personnel

Although the budget for waste management has increased dramatically in the past few years, it would be naive to suppose that lack of funding will not constrain future U.S. radioactive waste management programs. Considering the numerous competing demands on the federal budget and the strength of the political movement for lowering taxes, it is improbable that Congress and the president will appropriate full funding to develop the safest possible waste repository program. Although no mention may be made of them, especially in public, some trade-offs are likely between costs and safety in waste management as in nearly every other aspect of social life.

The increased funding to date, however, has brought larger waste management staffs, among other benefits. The NRC presently has over one hundred such personnel, whereas in early 1975 only one full-time professional and one clerical employee were assigned to the task. Yet, several qualitative problems with the personnel remain. First, they tend to have close ties to the nuclear industry; for example, one 1976 study found that 65 percent of top officials in the NRC were from businesses that had NRC contracts or licenses (Common Cause, 1976). There is presumably some danger that industry ties will reduce vigilance or cause blindspots in regulation. Moreover, some staff and operating procedures of the old AEC linger on. The NRC as an organization was patterned almost exactly like the AEC, four of its first five commissioners had ties to the AEC, and most of the early staff was simply transferred. Although this influence is now diluted, it is not altogether dissipated. Analogous problems exist at DOE.

Outlook on Technology

Opinions on nuclear waste disposal will be influenced in part by individuals' more general optimism or pessimism concerning "progress" through utilization of technology. A portion of the U.S. public has become skeptical of the claim that high technology contributes more to human and ecological well-being than it detracts. Such technological skeptics—rightly or wrongly—are likely to find fault with whatever waste management scheme is proposed.

Reinforced by impressive technical achievements such as the moon landing, however, a majority of the public remains at least moderately optimistic about progress through technology (Schon, 1967; LaPorte and Metlay, 1975). A vocal minority goes even further to maintain that almost any problem has a technical or scientific solution. Among this technological-fix group, there is a discernible technological or engineering mentality, particularly among scientifically trained individuals in government, industry, and academia (Wilkes and Gordon, 1977). The AEC and DOE have seemed to attract such people. One consequence is that they tend to distort and reduce complex social issues to narrower and more familiar technical ones. A specific effect of this tendency in the nuclear arena has been the neglect or gross underestimation of the probability of human error. This attitude has also led officials to ignore messy "political" tasks, such as gaining public acceptance, and to concentrate instead on technical achievements (Hoos, 1978; Metlay, 1978).

An unquestioning faith in high technology can also lead to overestimating the certainty of scientific knowledge (for example, concerning the stability of geological formations), as well as to underestimating the difficulties that occur during the process of transforming valid scientific theory into a workable, complex, new technology. For example, most nuclear engineers have long been confident that a safe nuclear waste repository can be constructed, and that the waste can be processed and transported safely. In the long term they may well be right. But the history of numerous technological developments suggests that malfunctions and surprises can be expected in the early stages. The Apollo program suffered a launch-pad fire that killed three astronauts; aircraft frequently have crashed because of simple, but unrecognized, design failures; the Bay Area Rapid Transit District continues to experience severe operating problems. In each of these cases, scientists and engineers had good reason to believe that the basis for success was available. Moreover, the designers had opportunities to build multiple prototypes and test models. Nevertheless, serious accidents and unanticipated problems occurred. There is no compelling reason to expect a nuclear waste repository to be any different.

Public Acceptance of the Waste Management Program

Technical containment of radioactivity is probably the most important component of the U.S. waste management program, but public acceptance of the program is nearly as important. This view has been reiterated by every recent federal report on the subject. In announcing his administration's policy on waste management in February, 1980, for example, President Carter made no fewer than twenty-five references to state, local, and public participation and to public acceptance of the program (Carter, 1980). A study sponsored by the NRC likewise asserted that "a correct decision reached in the wrong manner is unlikely to be accepted" by the public (Bishop, 1978, p. 55).

Unfortunately, it is not clear what the "right manner" of decision making on this issue would be. Of course, the closed style practiced in the 1950s and 1960s is unacceptable, but even open and pluralistic decisions do not guarantee public trust in and acceptance of the radioactive waste program. The open style of decision making characterizes most hot issues in American politics; but public trust in goverment nevertheless is very low (Ladd, 1982).

Something special appears to be needed if widespread public acceptance of the waste program is to be achieved. At the recommendation of the Interagency Review Group, the Carter administration staked its hopes for public acceptance on "the fullest possible disclosure to and participation by the public and the technical community" (Carter, 1980, p. 414). Two mechanisms were envisioned: mailing of a semifinal waste management plan to 15,000 people, who would be asked for written comments; and two-day regional meetings in five cities that would be open to anyone interested in attending (DOE, 1980).

There are several problems with such a plan. Only a tiny

Table 6.1 Attendance at Public Meetings on Radioactive Waste Management (Three 1977–78 EPA-Sponsored Forums)

Representatives of industry and utilities	200	27.3%
Employees of federal agencies	145	19.8%
University/research institute personnel	109	14.9%
Environmental groups	70	9.6%
State/local governments	56	7.7%
Public interest groups, foundations	43	5.9%
Press, publishers	18	2.5%
Unclassifiable and miscellaneous	11	1.5%
Private citizens	80	10.9%

SOURCE: Compiled from EPA 1977a, 1977b, 1978.

segment of the public will ever hear about it, and those attending the hearings will be highly unrepresentative of the general public, if past experience is a reliable guide (see table 6.1).[9] Participants are not asked focused questions on which clear answers can be given, and no attempt is made to measure participants' opinions systematically; as a result, published reports of past hearings have constituted little more than checklists of obvious concerns about radioactive waste. Finally, the hearings tend to focus on technical issues that are going to be decided by experts instead of on transcientific issues that will have to be resolved through the political process.[10] Studies of nuclear decision making in Europe suggest that public-relations efforts by other governments have been similarly flawed (Nelkin and Pollak, 1981; Nelkin and Fallows, 1978).

The idea that public acceptance of federal waste management policy could be built primarily through a process of hearings and mailings reflects an overly technical view of society: provide the facts, let anyone attempt to refute them, and then proceed. A broader, more sociological view of the task would

9. The NRC has requested funds to help defray the expenses that individuals incur in participating at NRC hearings on reactor licensing and other subjects, but has not received such funds to date.

10. See Chapter 7 for clarification of the nontechnical issues that might be suitable for public discussion and decision.

recognize the deep roots of public distrust concerning radioactive waste and that an extraordinary political effort may be required to overcome it. One possibility for such an effort is discussed in chapter seven (see also Woodhouse, 1982).

The Reagan administration appears to be going in the opposite direction, however, reducing emphasis even on public hearings and falling back almost entirely on the expectation that demonstrated technical prowess will lead to public acceptance of federal waste management policy (DOE, 1982).

Self-Interest as an Impediment to Wise Decisions

Short-range self-interest works in a variety of ways against the possibility of representative governments reaching workable decisions on technological issues. Elections are never more than two years away for a third of the Senate and for all of the House of Representatives. Neither business or political leaders nor the bulk of the general public typically stands to gain in the short-term from efforts to shape decisions that will prove effective over time. Such policies often tend to reduce employment or restrict business now (by prohibiting the production of a profitable but dangerous chemical, for example) in order to secure a public gain in the future. The heavily funded, well-organized corporations that would be restricted have a substantial, immediate interest in preventing the policy from being enacted. In contrast, typical members of the general public would realize only a small benefit many years in the future and lack the funds, time, or organization to learn about the issue and join a political battle (Olson, 1971).

Further biasing the outcome is the structure of the U.S. political system. There are many routes by which special interests can block political action that may be in the collective interest. If it ever gets out of committee, legislation has to pass not one house of Congress, but two; the courts can be used to blunt the effect of legislation; and regulatory agencies are easily "captured" by the industry that is supposed to be regulated (Bernstein, 1955; McConnell, 1970). Legislators elected without much assistance from their political parties are likely to avoid offending powerful interests who could aid an opponent at the next election. The fragmentation of the U.S. political system makes it more difficult than it is under a parliamentary system for citizens to know whom to blame when policies prove unwise.

Finally, the benefits and risks of nuclear energy and its wastes are difficult to apportion fairly. Since only a few repositories are necessary, some individuals can be expected to support nuclear electricity but to oppose repositories near them. Such individuals may tend to "buy" more nuclear waste and nuclear electricity than they would if the risks could be spread more evenly. Conversely, residents near the site of a proposed waste repository will have an incentive to oppose it, regardless of the overall national benefits. Senator John Stennis (D-Miss.), for example, has been an ardent proponent of military and civilian nuclear activities; but when DOE began looking at Mississippi salt domes as a potential site for a waste repository, the senator protested that he was "concerned about the people, the people. It's not just a little piece of land for storage that's involved here, but peace of mind for the people" (*New York Times,* 1982). Because the U.S. political system is organized geographically, localities with strong opposition to a repository stand an excellent chance of blocking its construction. If construction becomes politically difficult at all of the best geological sites, then a somewhat inferior site could have to be used. The net result (from the viewpoint of the nation) would be a lower quality nuclear waste disposal program.

Lack of Necessary Knowledge and Decision-Making Skills

Another obstacle to radioactive waste management is that most citizens lack the knowledge that would facilitate intelligent choices. Even many political leaders find technical information on nuclear energy incomprehensible. By sustained study, an intelligent person might learn enough to ask experts the right questions and spot obvious blunders. But relatively few people, even political leaders, are likely to be able and willing to devote the required time to more than a few issues—and a majority of voters do not study even a single policy problem in any depth.

Perhaps even more serious is the widespread lack of sufficient competence at the skills that facilitate effective decision making. Cognitive flexibility and capacity for abstract thought, open-mindedness, moral reasoning, interest and skill in reading, and other abilities are not successfully taught by most families, schools, or work organizations. Consequently, many individuals do not have the ability to think of public policy problems in more than short-range, personal, ad hominem, black-and-white terms: "The bastards in Washington/Middle East/big oil companies/environmental movement are going to raise my gas bill."

On some issues this lack of decision-making competence and knowledge may not be disabling to popular government. But on complex technological issues with partially irreversible consequences, citizen and leadership incompetence is a severe obstacle. For example, there is widespread ignorance and misunderstanding concerning the comparative risks, benefits, social consequences, and total financial costs of nuclear and other energy sources. Among other effects, this situation renders political participants susceptible to simplistic or distorted arguments and, therefore, reduces the probability of a high-quality decision. These limitations probably apply about equally to opponents and to advocates of nuclear power.

POSSIBLE IMPLICATIONS

Few of the biases and other obstacles to effective decision making discussed above are new, and most of them influence many different areas of public policy; however, these obstacles are especially important for nuclear waste and other high technology decisions. Historically, acceptability has been the primary criterion of whether or not representative governments were working well. Even incompetent, biased people and institutions can reach agreements that they will find temporarily acceptable. But problems of high technology tend to raise issues that require knowledgeable decisions that will prove durable over a long period of time. Even if everyone in the world agreed to dump raw nuclear waste into the oceans, for example, it might nevertheless be an unwise decision. Public acceptability is still important, of course, but it is no longer a sufficient criterion for political decision making.

Knowledgeable political decisions have become enormously more important in the United States in the past century. Because of population growth and economic change, political decisions now have a substantial impact on the daily life of millions of individuals and families in an interdependent urban society. In contrast, political decisions generally had a minor impact on daily life when a high percentage of colonial Americans were self-sufficient farmers.

Decisions also have to be more accurate to the extent that their consequences are potentially awesome, long lasting, and difficult to correct. It often takes many years for the consequences of technological decisions to appear—more than twenty-five years, for example, between the first production of DDT and the accumulation of persuasive evidence on its harmful ecological effects. Already the costs and risks of moving the radioactive wastes at Hanford are so great that the "temporary" storage there may have become irreversible.

For fully knowledgeable decisions about radioactive waste to have been initiated prior to its original generation in the 1940s, broad debate would have been required to compare the anticipated benefits with the costs and risks. Topics would have included alternatives to nuclear-generated electricity, dangers of nuclear theft and terrorism, the possibility of a reduction in citizens' ability to comprehend their society, growth of the federal government at the expense of states and localities, vulnerability to fanatical social-religious movements in the event of major panic from nuclear accidents, diminished prospects for democratic control, increased authority for experts, and other considerations that now are evident. Most of these issues still have received relatively little sustained consideration, and, in any case, the basic political choice has already been made. The U.S. has produced large quantities of electricity and nuclear weapons and will somehow, somewhere dispose of the resulting wastes at unknown risk. It is widely acknowledged that this basic choice was made without deep scrutiny by a relatively small number of political, economic, and scientific leaders; the public that acquiesced to the decision lacked the time, training, and political forum to weigh the options carefully.

This is not an indictment of nuclear decision makers in particular, since it is much easier to perceive these matters from hindsight and many other policy areas have had flawed decision processes. But the difficulties encountered, and still expected, in managing radioactive wastes might be used to examine whether the benefits of an increasingly technological society are worth the costs and risks. Is there a way to develop a reasonably comfortable lifestyle that would require less of the technical, social, and political complexity that makes public decisions on radioactive waste management so inherently difficult? Democracy has seldom been advanced as a method for making wise decisions. It has, instead, been admired principally as a good way of limiting gross abuses of governmental authority. The recent changes in the policy-making process for nuclear matters demonstrate again that representative government works for this purpose: the old nuclear subgovernment is gone. But the problem of radioactive waste remains, as do problems with toxic chemicals and other inherently difficult technology-based social problems. Even if we can assume that the technical aspects of radioactive waste management will be mastered, the remaining political obstacles are considerable. To depend on representative government, or perhaps any government, for accurate and far-sighted decisions on such matters is to create a burden for which government is not well suited.

REFERENCES

Bernstein, Marver. 1955. *Regulating Business by Independent Commission.* Princeton: Princeton University Press.

Bishop, William P. 1978. Observations and Impressions on the Nature of Radioactive Waste Management Problems. In Bishop et al., *Essays on Issues Relevant to the Regulation of Radioactive Waste Management,* pp. 51–59. Office of Nuclear Material Safety and Safeguards, U.S. Nuclear Regulatory Commission, Washington, D.C. NUREG-0412.

Boffey, Phillip M. 1975. *The Brain Bank of America: An Inquiry into the Politics of Science.* New York: McGraw-Hill, pp. 89–111.

Carter, Jimmy. 1980. President Carter's Message to Congress Outlining a National Radioactive Waste Management Program. Reprinted in *Congressional Quarterly Almanac* 36:38E–40E.

Cohen, Linda. 1979. Innovation and Atomic Energy: Nuclear Power Regulation, 1966–Present. *Law and Contemporary Problems* 43:67–97.

Colglazier, E. William, Jr. 1982. *The Politics of Nuclear Waste.* New York: Pergamon.

Committee on Interior and Insular Affairs, House of Representatives. 1978. *Nuclear Siting and Licensing Act of 1978, Part II.*

Committee on Waste Disposal. 1957. *The Disposal of Radioactive Waste on Land,* NRC no. 519, National Academy of Sciences-National Research Council, Washington, D.C.

Common Cause. 1976. *Serving Two Masters.* Washington, D.C.: Common Cause.

Deese, David A. 1982. A Cross-National Perspective on the Politics of Nuclear Waste. In Colglazier, 1982, pp. 63–97.

Del Sesto, Steven L. 1980. Nuclear Reactor Safety and the Role of the Congressman: A Content Analysis of Congressional Hearings. *The Journal of Politics* 42:227–41.

Department of Energy (DOE). 1980. National Plan for Radioactive Waste Management. Draft by Edward Mastal (mimeo).

———. 1982. *National Plan for Siting High-Level Radioactive Waste Repositories and Environmental Assessment.* DOE/NWTS-4 and DOE/EA-151. Columbus: Battelle Memorial Institute. Public Draft.

Environmental Protection Agency (EPA). 1977a. *Proceedings: A Workshop on Issues Pertinent to the Development of Environmental Protection Criteria for Radioactive Wastes, Reston, Virginia, February 3–5, 1977.* Office of Radiation Programs, Washington, D.C. ORP/CSD-77-1.

———. 1977b. *Proceedings: A Workshop on Policy and Technical Issues Pertinent to the Development of Environmental Protection Criteria for Radioactive Wastes, Albuquerque, New Mexico, April 12–14, 1977.* Office of Radiation Programs. ORP/CSD-77-2.

———. 1978. *Proceedings of a Public Forum on Environmental Protection Criteria for Radioactive Wastes.* Office of Radiation Programs. ORP/CSD-78-2.

Fallows, Susan. 1979. The Nuclear Waste Disposal Controversy. In *Controversy: Politics of Technical Decisions,* ed. Dorothy Nelkin, pp. 87–110. Beverly Hills: Sage.

General Accounting Office (GAO). 1968. *Observations Concerning the*

Management of High-Level Radioactive Waste Material. May 28. U.S. General Accounting Office, Washington, D.C.

______. 1974. *Isolating High-Level Waste from the Environment: Achievements, Problems, and Uncertainties.* December 18.

______. 1975. *The Liquid Metal Fast Breeder Reactor: Promises and Uncertainties.* OSP-76-1.

______. 1976. *Evaluation of the Publication and Distribution of "Shedding Light on Facts on Nuclear Energy," Energy Research and Development Administration.* EMD-76-12.

______. 1979. *The Nation's Nuclear Waste—Proposals for Organization and Siting.* June 21, EMD-79-77.

______. 1980a. *The Nuclear Regulatory Commission: More Aggressive Leadership Needed.* January 15. EMD-80-17.

______. 1980b. *The Energy Department's Office of Environment Does Not Have a Large Role in Decision-Making.* January 29. EMD-80-50.

Green, Harold P., and Zell, L. Marc. 1982. Federal-State Conflict in Nucler-Waste Management: The Legal Bases. In Colglazier, 1982, pp. 110–37.

Greenwood, Ted. 1982. Nuclear Waste Management in the United States. In Colglazier, 1982, pp. 1–62.

Hoos, Ida R. 1978. The Credibility Issue. In Bishop et al., pp. 20–30.

Hubbert, M. King. 1962. *Energy Resources: A Report to the Committee on Natural Resources of the National Academy of Sciences-National Research Council.* National Academy of Sciences-National Research Council, Washington, D.C.

Interagency Review Group (IRG). 1979. *Report to the President by the Interagency Group on Nuclear Waste Management.* U.S. Department of Energy, Washington, D.C. TID-29442.

Joint Committee on Atomic Energy (JCAE). 1966. *AEC Authorizing Legislation, Fiscal Year 1966, Hearings,* Statement of Dr. Glenn T. Seaborg, AEC Chairman, pp. 1374–76; and Appendix 5, An Analysis of Advanced Converters and Self Sustaining Breeders, pp. 1751–70. U.S. Congress, Washington, D.C.

Karpinski, Gene B. 1976. Federal Preemption of State Laws Controlling Nuclear Power. *Georgetown Law Journal,* 64:1323–41.

Ladd, Everett C., Jr. 1982. *Where Have All the Voters Gone?* 2nd ed. New York: Norton.

LaPorte, Todd R., and Metlay, Daniel. 1975. Public Attitudes toward Present and Future Technologies: Satisfactions and Apprehensions. *Social Studies of Science* 5:373–98.

Lewis, Richard S. 1972. *The Nuclear Power Rebellion: Citizens vs. The Atomic Industrial Establishment.* New York: Viking.

Lindblom, Charles E. 1977. *Politics and Markets: The World's Political-Economic Systems.* New York: Basic Books.

Lipschutz, Ronnie D. 1980. *Radioactive Waste: Politics, Technology, and Risk.* Cambridge, MA: Ballinger.

McConnell, Grant. 1970. *Private Power and American Democracy.* New York: Random House.

Metlay, Daniel S. 1978. History and Interpretation of Radioactive Waste Management in the United States. In Bishop et al., pp. 1–19.

Metzger, H. Peter. 1972. *The Atomic Establishment.* New York: Simon and Schuster.

Moss, Thomas H. 1982. What Happened to the IRG? Congressional and Executive Branch Factions in Nuclear Waste Management. In Colglazier, 1982, pp. 98–109.

Murphy, Andrew W., and LaPierre, D. Bruce. 1976. Nuclear "Moratorium" Legislation in the States and the Supremacy Clause: A Case of Express Preemption. *Columbia Law Review* 76:392–456.

Nelkin, Dorothy, and Fallows, Susan. 1978. The Evolution of the Nuclear Debate: The Role of Public Participation. *Annual Review of Energy* 3:275–312. Palo Alto: Annual Reviews, Inc.

Nelkin, Dorothy, and Pollak, Michael. 1981. *The Atom Besieged: Extraparliamentary Dissent in France and Germany.* Cambridge: MIT Press.

New York Times. 1975. Atomic Industry to Promote Views. January 17, p. 34.

______. 1976. Letters to the Editor. December 2, p. 42.

______. 1979. How National Academy of Science Decided to Halt a Nuclear Waste Report Is Disputed. June 25, p. 13.

______. 1980a. 25,000 in Washington Hold Antinuclear Rally in Rain. April 27, p. 26.

______. 1980b. Maine to Vote Tuesday on Closing of Nuclear Plant. September 21, p. 28.

______. 1980c. Nuclear Safety Report Criticizes Industry and Regulatory Agency. November 26, p. 15.

______. 1982. Senators Debate Atomic Waste Bill. April 29, p. 20.

Nuclear Regulatory Commission (NRC). 1975. *Reactor Safety Study* (Rasmussen Report). Washington, D.C. WASH-1400 or NUREG-75/104.

______. 1977. *Workshops for State Review of Site Suitability Criteria for High-Level Radioactive Waste Repositories.* Office of Nuclear Material Safety and Safeguards. NUREG-0353 (October).

Olson, Mancur, Jr. 1971. *The Logic of Collective Action.* Revised edition. Cambridge: Harvard University Press.

President's Commission on the Accident at Three Mile Island. 1979. *Report of the President's Commission on the Accident at Three Mile Island.* Washington, D.C.: U.S. Government Printing Office.

Schon, Donald A. 1967. *Technology and Change: The New Heraclitus.* New York: Delacorte Press.

Science. 1974. Low Marks for the AEC's Breeder Reactor Study. Vol. 184, p. 877.

Sills, David L.; Wolf, C. P.; and Shelanski, Vivien B. 1982. *Accident at Three Mile Island: The Human Dimensions.* Boulder: Westview Press.

State Planning Council on Radioactive Waste Management. 1981. *Report to the President by the State Planning Council on Radioactive Waste Management.* State Planning Council on Radioactive Waste Management, Washington, D.C.

Study Group on Light Water Reactor Safety. 1975. Report to the APS by the Study Group on Light Water Reactor Safety. *Reviews of Modern Physics,* vol. 47, supp. no. 1.

Temples, James R. 1980. The Politics of Nuclear Power: A Subgovernment in Transition. *Political Science Quarterly* 95:239–60.

U.S. District Court. 1979. Pacific Legal Foundation v. State Energy Resources Conservation and Development Commission. 472 F. Supp., pp. 191–201.

______. 1980. Pacific Gas and Electric Company v. State Energy Resources Conservation and Development Commission. 489 F. Supp., pp. 699–704.

______. 1981. Washington State Building and Construction Trades Council v. Spellman. 518 F. Supp., pp. 928–35.

U.S. Supreme Court. 1976. Train v. Colorado Public Interest Research Group, Inc. 426 U.S., pp. 1–25.

Varanini, Emilio E., III. 1982. Consultation and Concurrence: Process or Substance. In Colglazier, 1982, pp. 138–59.

von Hippel, Frank. 1977. Looking Back on the Rasmussen Report. *Bulletin of the Atomic Scientists* 33:42–47.

Wall Street Journal. 1971. Atom Age Trash: Finding Places to Put Nuclear Waste Proves a Frightful Problem. January 25, p. 1.

Wilkes, John M., and Gordon, Gerald A. 1977. Technology, Technologists and Social Problems. Presented at the annual meeting of the American Sociological Association.

Woodhouse, Edward J. 1982. Managing Nuclear Wastes: Let the Public Speak. *Technology Review,* October, pp. 12–13.

Zinberg, Dorothy S. 1982. Public Participation: U.S. and European Perspectives. In Colglazier, 1982, pp. 160–87.

CHARLES A. WALKER, LEROY C. GOULD, AND EDWARD J. WOODHOUSE

7 VALUE ISSUES IN RADIOACTIVE WASTE MANAGEMENT

A central question underlies the first six chapters of this book: can radioactive wastes be managed safely? To this question there is no simple answer, not just because of the technological complexity of the problem, but also because of an ambiguity in the question itself—how safe, after all, is "safe"? If "safe" means absolutely no risk, then it can be said with certainty that the answer to the question is no; radioactive wastes cannot be managed with zero risk either to workers or the public. If, however, "safe" means "reasonably safe" or "safe enough," then the question simply has been begged—how safe is "reasonable" or "enough"?

William Lowrance, also concerned about this ambiguity in the concept "safe," has suggested (1976, p. 8) that safety decisions resolve into two components: measuring risks, that is, predicting the probabilities and consequences of events, and judging safety, that is, judging the acceptability of that risk. The first component, he contends, is or at least ought to be an objective, scientific pursuit while the second is a matter of personal and social value judgment.

> Failure to appreciate how safety determinations resolve into the two discrete activities is at the root of many misunderstandings. In one of the most common instances, it gives rise to the false expectation that scientists can *measure* whether something is safe. They cannot, of course, because the methods of the physical and biological sciences can assess only the probabilities and consequences of events, not their value to people. Scientists are prepared principally to measure risks. Deciding whether people, with all their peculiarities of need, taste, tol-

erance, and adventurousness, might be or should be willing to bear the estimated risks is a value judgment that scientists are little better qualified to make than anyone else. [Lowrance, 1976, p. 9]

In the case of radioactive waste management, then, questions such as whether glass or ceramics will maintain integrity longer in a particular geological environment, or whether salt or granite is the better barrier to groundwater intrusion, are technical issues; the only "safety" questions involved are the probabilities that one or another of these barriers will fail. Whether radioactive wastes, no matter what their form or their host rock, should be placed in permanent disposal or retrievable storage, however, is a question of value judgment involving trade-offs between such things as short-term and long-term safety, costs, peace of mind, and institutional stability.

It is precisely because value judgments are involved that the public has a legitimate role to play in making decisions about radioactive waste management. This role will not be played well, however, in our opinion, unless the ongoing debate about radioactive waste management is clarified and technical issues involving risk measurement are separated from value issues involving the acceptability of risk. Such a reorientation will not be easy but should not be impossible. The remainder of this chapter is one attempt to identify and comment on some of the more important value issues in radioactive waste management.

PUBLIC INTEREST IN TECHNICAL ISSUES

As noted in Chapter 2, many technical problems remain to be solved before efficient systems for the management of radioactive wastes can be designed, constructed, and operated. Research and development in this field have begun to receive adequate attention and funding only in very recent years, but the record to date is encouraging in that several difficult problems have been solved and there are indications that solutions to the remaining technical problems can be developed. The record to date in the management of radioactive wastes from mining, milling, conversion, enrichment, fuel fabrication, reactor operation, and military-related processes is marred by instances of tank leakage and other accidents and errors. This record is also basically encouraging, however, in view of the magnitude of operations and the complexity and unique features of nuclear fission processes. We see no reason to suppose that technically efficient solutions to problems of radioactive waste management cannot be developed.

The search for technically efficient solutions will involve numerous points where decisions about risks and about values will have to be made. In this section we attempt to identify some of these points of conjunction of technology, risks, and values, and we suggest that these points create the issues of most direct concern to the public.

The Future of Commercial Nuclear Power

Foremost among the issues of nuclear energy policy is the question of whether this nation should commit current and future generations to nuclear energy for a large share of their energy needs. Arriving at an answer to this question will require considerations that go far beyond the subject matter of this book, including assessment of risks in other parts of the nuclear fuel cycle, comparisons with alternative energy sources, and the role of conservation in our energy future as well as numerous social, political, and moral issues. A successful radioactive waste management policy is a necessary condition for continuation of the development of nuclear energy, but it is not sufficient to justify such development.

Even if a decision to abandon nuclear power were made, however, there would remain the problem of storing or disposing of the considerable quantities of radioactive wastes now in existence. The magnitude of the problems will increase significantly if more nuclear power plants are built and operated, of course, but the current backlog alone is enough to require that policy decisions be made and that a "safe" plan for radioactive waste management be developed. Whether or not we develop a radioactive waste management program is not a value issue to be settled: we need one.

Radiation Standards

The basic worry about radioactive waste management systems is that the operation of such systems will result in some increase, above background levels, in exposure of humans and other components of the biosphere to alpha particles, beta particles, gamma rays, and neutrons. In view of the fact that irradiation is known to result in increased incidence of cancer and genetic defects, decisions about allowable dosages to workers and members of the public are of direct concern to the public. The setting of standards is, in a very real sense, a trade-off between costs and people, since more rigid standards will cost more to achieve but will result in saving some lives and in some improvement in the quality of life.[1]

The design and operation of radioactive waste management systems will depend heavily on specifications of allowable radiation doses to workers and the general public. Although radiation standards already exist for the nuclear industry and other users of radioactive materials, these standards, as noted in Chapter 3, are based on imperfect knowledge of the effects of low-level exposures on living organisms and could easily change as more accurate data become available.

The radiation standards set for radioactive waste management need not, however, be the same as those that are established for operating nuclear power plants, since what is allowable in each case needs to be based not only on the risks but also on the benefits of that activity. As noted in Chapter 5, the benefits of such things as nuclear medicine, scientific research, nuclear power, and radioactive waste management are likely to be viewed quite differently, both by members of the general public and by professional risk managers, and thus the judgments of what risks are acceptable in each case are likely to differ considerably. Those people who were polled on the acceptability of radioactive waste management risks (Chapter 4), for

1. Such trade-offs are also involved in making decisions about air traffic control, automobile design, production and use of coal, and a variety of other activities in modern life. They are not unique to nuclear power.

example, indicated, on average, that one death per year from radioactive waste management activities would be acceptable in the short run and one death per hundred years would be acceptable in the long run. Complying with such stringent standards would be very expensive.[2]

Decisions about radiation standards in radioactive waste management thus depend primarily on values and involve trade-offs between risks and benefits and on distributions of risks over current and future generations. Radiation standards are the basic issue in radioactive waste management. They are of primary importance to workers, to persons living near nuclear facilities of all kinds, and to other members of the public. Decisions about all other issues, including those outlined below, must be approached with questions in mind about the extent of radiation exposure due to each technological alternative and the effect of this exposure on living organisms.

Some members of the public will be interested not only in the setting of radiation standards but also in methods used for radiation monitoring of nuclear facilities, the results of such monitoring, and interpretation of the results in terms of probable health effects.[3] That is, they will want to know whether the standards are being met. In this part of the complex task of managing radioactive wastes public perception of the credibility of the agencies responsible for monitoring will be particularly important.

Storage or Disposal

Choices between storage and disposal of radioactive wastes will necessarily be made partly on the basis of costs and available

2. Note, however, that the same research subjects said that they would be willing to pay $5 per month per family to meet these standards. On a national basis this would be about $5 billion per year if every family in the United States paid it or about $500 million per year if apportioned according to the percentage of electricity generated by nuclear reactors. As noted in Chapter 2, a preliminary estimate of the cost of disposing of commercial spent fuel is $232 per kilogram, or about $350 million per year, at present levels.

3. Current nuclear facilities are subject to careful monitoring, of course, and data on radiation levels and predictions of health effects are available in government publications. For example, see ERDA-1538 for data on the Hanford Reservation operations.

technology, but these choices are also dependent on values in several important respects.[4] In the following discussion we use the most significant and problematic waste from commercial operations, spent fuel, as an example, noting however that choices between storage and disposal must be made for other types of radioactive waste.

Almost all of the spent fuel generated by nuclear power plants in this country is currently in storage in water-cooled basins at reactor sites, a technology that has proven so far to be safe and relatively inexpensive. It is likely that this practice will continue for the next decade or more while decisions about reprocessing and various schemes for continuing storage or disposal are being made. Such continuation has advantages other than simply buying time. Longer storage times at reactor sites result in simplification of future problems of transportation, storage, and disposal simply because levels of radioactivity and rates of heat generation decrease with time. Also, spent fuel is contained in engineered, accessible structures where control over containment, heat generation, and avoidance of criticality is easily provided, where changes in waste management techniques can be made if required, and where the spent fuel is easily retrievable for further processing. Furthermore, storage at reactor sites keeps the benefits and risks of nuclear power to some extent coterminous since those who use the electricity live in the region where a plant is located, although there are inequities in risks between those who live very close to a plant and those who live farther away.

Continued storage of spent fuel at reactor sites has several disadvantages, including the fear that radioactive wastes stored at or near the surface of the earth would be more likely to migrate into the biosphere than would wastes in deep repositories. Migration can occur as a result of inadequacies in design and operation, such as leaks and unanticipated accidents and errors, natural events, such as earthquakes, or deliberate ac-

4. Storage is the placement of wastes in above-ground structures or relatively shallow caverns to which there are continuing means of access and methods and equipment for shifting the position of the wastes or removing them to other locations. Disposal is the placement of wastes in relatively deep underground formations with provision for ultimate backfilling and sealing.

tions. These possibilities suggest that monitoring and surveillance requirements are severe for any case where radioactive wastes are placed in storage near the earth's surface.

The size of the population placed at risk by on-site storage is large in those instances where power plants are located near large population centers. The size of the population at risk can be reduced by providing storage in off-site basins (away-from-reactor sites, or AFRs, in government terminology) in sparsely populated areas. In fact, some utilities are having difficulties in providing enough on-site storage space, and AFRs might become necessary as more spent fuel accumulates. One result of such a development would be to place at risk populations not receiving the benefits of electricity; that is, those living along transportation corridors and near AFRs.

The concept of away-from-reactor storage can be extended to become a long-term solution to the problem of what to do with spent fuel. After a decay period of several years in water-cooled basins, spent fuel could be packaged and transferred to simple air-cooled concrete silos located in remote areas. This would provide continued accessibility, retrievability, and control, but it would leave the wastes in relatively exposed conditions and would require long-term monitoring and surveillance.

Disposal of high-level wastes in backfilled and sealed geologic formations is the basis of much current planning for radioactive waste management. This procedure offers the significant advantages of reducing the risk that radioactive waste would migrate into the biosphere and minimizing requirements for monitoring and surveillance. The seeming finality of disposal could be realized, however, only with a loss of opportunities for retrieval and for control in case something goes wrong. For these reasons, it is essential that disposal of spent fuel be practiced only after considerably more research and development are done and only if it is clear that the fertile and fissile elements in spent fuel will not be useful to current or future generations.

Storage and disposal thus differ in terms of the quantity of radioactivity that might enter the biosphere, the distribution of risks and responsibilities over segments of current and future populations, vulnerability to natural events and human actions, retrievability, and the degree of control available, all of which involve value judgments. It seems to us that debates over radioactive waste management policies have focused too sharply on disposal in geologic formations and given too little attention to possibilities for interim or long-term storage. Storage will be, after all, the only method available for managing high-level radioactive wastes for a decade or more. Furthermore, even if disposal plans were fully developed and implemented, a large inventory of wastes would continue to be stored at the surface while decaying, being prepared for disposal, and being transported. For these reasons storage facilities and procedures deserve more discussion.

Reprocessing

One of the major decisions to be made in radioactive waste management is whether or not to reprocess spent fuel rods. Even if recycling uranium and plutonium were not at issue, which in fact it is, separating spent fuels into their constituent parts—uranium, plutonium, hulls and hardware, and fission products—has clear advantages. Using such a scheme, each waste form could be managed in optimum fashion. In one possible scenario, for example, plutonium and uranium would be stockpiled, and hulls and hardware and fission products would be disposed of with little concern that they might someday be regarded as valuable resources. This scenario would have two important advantages for future generations: it would save the potentially useful fissile elements for future use as energy sources, and it would render the undesirable products more suitable for safe, permanent disposal.

Reprocessing, however, also has several disadvantages. First of all it would increase supplies of materials from which nuclear weapons could be fabricated, thus increasing risks of nuclear proliferation; and second, it would increase the risks of accidents due to handling since reprocessing would place a new, complicated manufacturing step in the management scheme and would probably also increase transportation risks. Although these risks would be borne primarily by present generations, who are benefitting from nuclear electric power, those at greatest risk would not necessarily or even probably be those who are receiving the greatest benefits, assuming that reprocessing plants would be located in areas of low population density rather than in areas close to presently operating nuclear plants. Several value judgments, in other words, will be required in deciding whether or not to reprocess spent fuel.

Site Selection

Given either spent fuel or high-level waste products from reprocessing, the next question to be considered is where the fuel assemblies or other high-level wastes should be disposed of and how many disposal sites there should be. As indicated in Chapter 1, present planning indicates that an immediate mandate to build a waste repository would force a decision to locate that repository in a salt formation somewhere in the continental United States. With more time for research, other rock formations could be considered, as could disposal beneath the oceans. The risks of proceeding immediately to build a repository, then, are the risks that on-shore salt formations might not be the best choice, in the long run, for repository sites. Assuming that disposal in on-shore salt formations is probably safer in the short run than surface storage, however, this option would probably have the effect of lowering risks to present generations but possibly increasing risks to future generations. Given the public opinion data reported in Chapter 4 that indicates that Americans may favor minimizing risks to future, rather than present, generations, this option might not be favored by a majority of Americans.

Site selection also involves decisions about the number of repository sites: wastes can be concentrated in one or a very few sites, or they can be distributed over a large number of sites. The use of fewer sites leads to more complicated transportation problems but simplifies monitoring and surveillance requirements. Multiple siting reduces the magnitude of effects of unanticipated events, since less materials would be located at each site, but monitoring and surveillance would be more difficult. Although one analyst (Rochlin, 1976) has argued that multiple siting is safest overall, the issue remains one of value judgments that will have to be made before final site selection can be made.

No matter whether one or several sites are to be chosen, however, the question of location remains. Here again, as with location of surface storage or reprocessing facilities, the questions emerge of whether these sites should be in unpopulated regions of the country, which would reduce the overall risk to the entire population, or near presently operating nuclear power plants. Although the latter strategy would keep the risks and benefits of nuclear power more coterminous, it would also increase the total risk of exposure to populations and for this reason probably will be unattractive in the final analysis. Given furthermore that geological considerations will of necessity play a major role in site selection, questions of location will ultimately involve technical considerations that could well overshadow ethical or political considerations. Nevertheless, many, if not most, people are going to want to know why a facility should be located in their backyard rather than somewhere else. Site selection, therefore, could prove to be one of the most contentious of all issues facing radioactive waste management.

Decontamination and Decommissioning

Problems of managing radioactive wastes will become considerably more visible to the public when current commercial power reactors reach the end of their useful lives and must be decontaminated and decommissioned. While most of these problems are at least two or three decades in the future, some experience is available from decontamination and decommissioning of a few military and experimental reactors. The removal of reactor cores and contaminated materials and equipment from power plants must be carefully planned and executed, but it can be done. Such removal would leave massive structures with enough residual radioactivity that they would have to be entombed, and careful monitoring and surveillance would be required for several decades. The presence of such a structure could have significant psychological and economic implications for people in surrounding areas, and these might be important enough to require dismantling and transport of rubble to disposal sites. Public interest in this phase of radioactive waste management is likely to be high, and public opinion and values will have a strong influence in the choice of methods to be used.

Decontamination and decommissioning of other facilities in the nuclear fuel cycle will also be necessary. Reprocessing plants, if they are built and operated, will pose unique and difficult problems because of liquid wastes. A case in point is the abandoned Nuclear Fuel Services reprocessing plant in West Valley, New York, where cleanup costs are estimated to be several times the cost of building the plant. Again, public interest in the methods to be used for decontamination and decommissioning of such plants will be high, particularly among people living in the vicinity of the facilities.

Summary

When schemes or facilities for radioactive waste management are proposed for public consideration we suggest that the kinds of information that will be of greatest interest include the following:

1. A statement of the maximum allowable radiation dosages that serve as a basis for design and estimates of probable health effects of these maximum dosages.
2. Estimates of the actual radiation dosages that will be experienced and of their probable health effects.
3. Analyses of vulnerability to natural events and human actions.
4. Descriptions of provisions for control of materials and radioactivity during routine operations or emergencies.
5. Descriptions of provisions for radiation monitoring of facilities and processes, assignment of responsibilities for monitoring, and plans for making available to the public information on the results of monitoring and estimates of probable health effects.
6. Analyses of the distribution of risks over segments of current and future populations.
7. Analyses of the distribution of responsibilities over segments of current and future populations.

We suggest further that these kinds of information are needed for assisting the public in understanding and commenting on decisions about choices between storage and disposal, choices of sites for facilities, choices between reprocessing and storage or disposal of spent fuel, and choices among methods for decontamination and decommissioning of facilities.

COST CONSIDERATIONS

Chapter 2 contains estimates of the costs of disposing of radioactive wastes from commercial and military wastes and a cautionary note that these estimates might be appreciably lower than the actual costs that will be incurred. But whatever the costs, an important value question is, Who pays? The answer to this question for military wastes is straightforward—the costs will be paid by the federal government as part of the defense budget; that is, the American taxpayer will pay.

As to commercial wastes, the Interagency Review Group (IRG, 1979) concluded that "appropriate cost of storage and disposal of any waste generated in the private sector should be paid for by the generator and borne by the beneficiary." Before this could be done by a one-time charge, however, costs of current and future storage and disposal technologies would have to be known with much greater accuracy than they are now. The one-time charge to utilities is inappropriate since an estimate lower than actual costs would require that persons other than the beneficiaries bear the extra cost. An initial charge supplemented by periodic assessments would be more equitable. Even this plan has limitations, however, since much of the cost will be borne a decade or more after spent fuel is removed from a reactor, and the continued solvency of a utility company is by no means guaranteed.

An even more difficult problem in equity is posed by the costs of decontamination and decommissioning of commercial nuclear power facilities. Here again the costs are uncertain since limited experience with this stage of the nuclear fuel cycle is available. Furthermore, it is not known whether less costly alternatives, such as mothballing, will prove acceptable or whether plants will have to be dismantled and placed in repositories. If the public demands dismantling, costs will be much higher. In either event, decontamination and decommissioning costs will come due four or more decades after the initial operations of a power plant, creating potential intergenerational problems.

The management of radioactive wastes raises some interesting compensation issues, but we see only one significant difference, to be discussed later, between compensation issues raised by radioactivity and those raised by other forms of environmental pollution. One of these issues is the matter of compensation for damages when they result from routine operation of facilities or from accidents and errors in either design or operation; similar issues are raised in the operation of nuclear power plants, coal-fired power plants, and other industrial facilities. The general problems of compensation for environmentally-mediated damages to persons and property are difficult ones, particularly in view of the fact that it is frequently not possible to prove that particular damages result from the operation of a particular facility. Thus a person suffering from respiratory ailments cannot usually prove that his or her condition is due in significant part to the operation of a nearby coal-fired power plant. Similarly, it would be very difficult to prove that a particular case of cancer or a genetic defect is due to radiation from a particular source. The problems are rendered even more difficult by the fact that health problems are sometimes not apparent until many years after exposure to environmental pollution. These issues of compensation for damages due to environmental pollution are concerns of economists, government officials, and others. We pursue them no further here simply because of space limitations and because they represent a much broader issue than the topics we are addressing.

There is, however, one unique compensation issue in radioactive waste management, and that is the possibility of catastrophic damage. While the probability that a waste storage site will achieve criticality and generate unmanageable quantities of heat or that large areas will be rendered uninhabitable because of transport of wastes by water is very low, these events are possible. This leads to a situation similar to the problem of liability insurance for companies operating nuclear power plants. In that case liability insurance up to $140 million is provided by commercial firms, and additional insurance up to a total of $560 million is provided by the government under the Price-Anderson Act and paid for by the owners of power plants. A problem is created by the possibility that a nuclear accident could result in damages far exceeding $560 million. Some writers (Duderstadt and Kikuchi, 1979) note that damages exceeding this maximum would result in provision of additional relief by the federal government. Others (Shrader-Frechette, 1980) are concerned that such additional relief for victims might be available only after a slow and tedious legal process or not at all unless the proof of damages is clear.

DECISION-MAKING PROCESSES FOR RADIOACTIVE WASTE MANAGEMENT

It should be clear by this point that numerous value judgments will have to be made in further developing the U.S. program for managing radioactive wastes. On none of the substantive issues discussed above do we offer an answer. There are no sure answers, only informed preferences. It is curious, therefore, that the Department of Energy and its predecessor agencies have approached waste management as if the significant questions were technical ones that could be decided largely by experts in and around DOE. In our judgment, correcting this error is the single most important task facing the U.S. radioactive waste management program. This concluding section explains why we believe this, and it offers suggestions on how to begin correcting the mistaken technical emphasis.

The Technically Oriented View

Although somewhat oversimplified, the technically oriented view of waste management might be stated in the following terms. The goal of waste management is to provide containment of radioactivity at an acceptable cost. This is a complex technical problem that can best be solved by leaving decisions largely in the hands of nuclear engineers and other experts. To the extent that disagreement exists among the responsible technologists, that disagreement will substantially dissolve with further research and discussion. The nuclear community, including DOE, must persuade the public to take a rational view of the issue. Rapid progress in permanent waste disposal is the best method of persuasion, and appropriate techniques are either at hand or will be developed in the near future.

This view is not unreasonable. As the basic emphasis of a U.S. nuclear policy, however, we believe that the technical view is too narrow. The first problem is the presumption that the main issues in waste management are technical. The preceding portions of this chapter demonstrate, we hope, that value issues are at least as central as technical ones. Indeed, if we accept the assurances of the nuclear community that waste management is technically feasible, then the problematic issues are overwhelmingly social and political.

Second, because so many of the basic issues in waste management involve conflicts of values, significant areas of uncertainty and disagreement will surely remain even after several more decades of research, development, and political conflict. Highly respected scientists and vocal activists from industry or environmental groups will oppose whatever choices are made for waste management. Hence there is virtually no chance of achieving public confidence through the melting away of opposition. The apparent reliance of the AEC, ERDA, and now DOE on achieving widespread agreement on the technical components of waste management has been an illusory hope.

Third, slightly more plausible has been the hope that quick burial of wastes in a repository might demonstrate that concerns over nuclear waste are not worth public attention. The out-of-sight, out-of-mind syndrome is real, but weighing against this strategy are the difficulties and ambiguities in any such demonstration of waste-containment technology. Some portions of the technology—transportation, conversion of wastes to forms suitable for storage or disposal, cladding, placement of wastes to provide for short-term containment and heat removal and avoidance of criticality—hardly need to be demonstrated. Another portion of the technology—long-term containment—cannot be demonstrated over a short period, nor can it be demonstrated that monitoring, surveillance, and management institutions will function successfully in the long term. It is probable, moreover, that a new and complex technology will encounter problems in the early stages that will shake rather than bolster public confidence. Demonstration projects will be useful in providing basic technical and economic data for selecting among alternative technologies, but they are not likely to result in near-term improvements in public confidence.

A More Value-Oriented View of Waste Management Decision Making

On technical issues, the standard by which a decision-making process can be evaluated is the technical correctness of the outcome. On value issues, a more appropriate standard is the acceptability of the outcome. A value-oriented view of the U.S. waste management program would give heavy emphasis to value issues, would take great pains to build public confidence in the waste management program, and would construct a decision-making process that anticipates continuing disagreement.

What sort of nuclear waste decision-making process might be able to achieve these goals? We believe that the key lies in a conspicuous break with past practices, which both those who participate most closely in waste management decisions and the broader public should recognize as something new and commendable. Three controversial sets of actions deserve consideration.

Putting the Right Federal Agency in Charge. Can the Department of Energy transform its narrow technical approach to the waste program into a broader, value-oriented view? If not, perhaps responsibility for managing the program should be transferred.

As described in Chapter 6, although no one agency or department has been responsible for the overall development of a national radioactive waste management program, the Department of Energy has played the most important role to date.

The Interagency Review Group recommended in its 1979 report that DOE be officially given primary responsibility. Most of those who commented on the IRG report agreed that there should be a single lead agency, but several questioned whether DOE was the appropriate choice. Some argued that "DOE would continue the policies of its predecessor agencies which were judged to be inadequate," while others held that "DOE is more committed to disposing of waste quickly than carefully."

Numerous observers have perceived DOE as caught in a conflict of responsibilities because the department both promotes nuclear power and disposes of nuclear wastes. Congress turned regulatory authority of power plants over to the Nuclear Regulatory Commission to avoid a conflict in that area. A similar step is recommended for waste management by at least one group critical of federal waste management policy, the Union of Concerned Scientists. They argue for establishing an entirely new authority, independent of all existing federal agencies and with no ties to past failures in waste management. The new authority would assume responsibility for all phases of nuclear waste, from research to policy making to daily management. The director of the new authority, the Union advocates, "should be a distinguished individual with no previous ties to the nation's nuclear program who can restore a much-needed credibility to the radioactive waste management program" (Lipschutz, 1980, p. 173).

There are obvious advantages to such a proposal. But there are serious drawbacks as well. First, it is difficult to start a federal agency from scratch. Many personnel from DOE and other existing agencies would be needed for their experience and knowledge. Second, the waste program will require resources and coordination that a new, small agency would find hard to muster. Third, as the IRG asserted, "any transfer to a new agency would involve considerable delays and disruption of on-going programs and would not, in itself, necessarily solve the problems perceived to exist with DOE and its predecessor agency programs." Fourth, political scientists paint a sorry picture of the normal history of "independent" regulatory commissions; they seldom work out as planned. Other difficulties could be added.

So in terms of efficient progress toward developing technical procedures for managing nuclear waste, the proposed transfer of responsibility might be as much a liability as an asset. Perhaps new personnel at the top of DOE under the Reagan administration (or its successor) will bring a change away from a narrowly technical focus. If not, then even the slowdown accompanying creation of a new agency might be warranted if that is the price of obtaining federal leadership that takes a more value-oriented approach to waste management.

Facilitating Debate. Government agencies in the U.S. are not good at persuading the public of much of anything, certainly not on issues as controversial as nuclear waste. Is it perhaps a mistake for DOE to even attempt persuasion? A federal agency that wanted to maintain its own legitimacy, and to foster the legitimacy of the government more generally, would inquire what the public wants and would facilitate debate on value-oriented aspects of radioactive waste management.

The inquiry function is now performed largely through public hearings and workshops. As discussed in Chapter 6, these activities have been virtually token efforts that could not possibly accomplish much. If the public is to be genuinely brought into the radioactive waste decision process, substantial changes would be necessary. The audience for hearings and workshops should be a random cross-section of the public, not a self-selected group. In order to achieve this, it would be necessary to pay all expenses of participants and probably a substantial honorarium also. In place of the amateurish, technical focus of many current hearings, DOE would have to employ professionals capable of translating complex issues in waste management into simpler terms, with a focus on values rather than on technical problems. To measure participants' preferences, expert polling could provide a statistically valid picture in considerable depth, much superior to the checklist of concerns produced in summaries of current public hearings.

Planning and conducting public hearings and workshops based on this model would be an ambitious and expensive undertaking. Whether the results of such a procedure would be acceptable to a large segment of the public as guidelines for waste management programs is uncertain. Yet, if the purpose of public hearings and workshops is to determine how a cross-section of the public reacts to carefully explained waste management plans, some such procedure must be devised to overcome the serious deficiencies of hearings and workshops as currently conducted. We suggest that the radioactive waste management problem has some characteristics that make it suitable for some interesting experiments on this model. It is sufficiently complicated technically that it can serve as a representative of a broad class of problems in which technology, risks, and value judgments are involved. At the same time it is not so complicated but that the technological alternatives, risks, and value judgments can be explained to a general audience. Furthermore, a credible interim strategy—continued storage at reactor sites or AFRs—exists and could be used to conduct experiments on long-term public participation that might yield valuable suggestions and provide a guide for making decisions about other technologies in the future.

Debate on a much broader scale could be facilitated by improvements in the dissemination of information about radioactive waste management. A large amount of information is currently available, of course, through printed materials of all kinds and radio and television broadcasts, but much of this information is too technical, unnecessarily detailed, or focused on issues of little interest to the public; much is so one-sided as to lose its credibility. These faults in information are not surprising in a technology that is relatively recent and characterized by a kind of risk, radiation exposure, that is not familiar to most people. The quality and relevance of information is improving, but further improvements will be needed if members of the public are to grasp the issues and become effective participants in the decision-making process through public hearings and workshops or, as suggested in the next section, through referenda.

Finding a Decision Process that Losers Will Accept. No matter how thorough the debate, no matter how persuasive the majority's case, there will remain a great many people who believe that the wrong decisions have been made about nuclear waste—even if they do not know what the "right" decisions are. Is it possible to structure a decision process that the losers, and future generations, will find acceptable even if they do not like the substance of the decisions?

As discussed in Chapter 5, people are much more willing to accept voluntary risks that they have chosen than involuntary ones that have been thrust upon them. Even people who vote against the winner in elections tend to feel that they have been consulted and to accept the victor as the legitimate holder of the office for the prescribed term. Antinuclear dissenters in the states that have held nuclear referenda have tended to accept the majority's decision and seem to have muted their subsequent opposition. The same has occurred in Sweden, where widespread public participation has helped decide the future of nuclear power.

What this implies for a decision process concerning radioactive waste management is that the current generation must feel that it has been consulted. We doubt that any decision made by DOE and its nuclear consultants will leave the public feeling consulted. Similarly, given the low levels of confidence typically expressed by the public in Congress and the presidency, it is doubtful that waste decisions made through the normal political process will be considered fully legitimate. So although it would be difficult and unprecedented, we believe serious consideration should be given to a series of national, regional, and local referenda on radioactive waste management. DOE, Congress, and the president might agree on a technically acceptable plan and put it to the public for acceptance or rejection; there could be either a single vote on the entire package or a separate vote on each aspect of it. Alternatively, two or more technically acceptable options could be constructed for each major value choice, and the public asked to make a series of value choices that cumulatively would outline the U.S. strategy for waste management.

Many objections can be raised against this idea. The reasons typically given for excluding the public from decision-making processes are incompetence and unwieldiness (Dahl, 1970). This chapter, however, constitutes an extended argument that the main choices facing the U.S. in waste management are largely value disputes in which competence is a relatively minor consideration. The best we can hope for is a publicly acceptable program that is engineered competently. So while public ignorance about radioactive waste (discussed in Chapter 4) is nothing to rejoice about and should be greatly reduced through conspicuous and extended debate, neither must public ignorance necessarily bar participation in the value choices that are impending. It is worth noting, moreover, that referenda have been held in many states and cities for years, some on very complex issues. While voters sometimes make decisions that are contrary to the bulk of expert opinion, such as in California's Proposition 13 that drastically cut property taxes, no one has demonstrated that the average effects of such referenda are any worse than decisions made by elected and appointed officials.

But would referenda be too time consuming, costly, and otherwise unwieldy? The matter has not received sufficient detailed analysis to offer a solid answer, but part of the answer would turn on how valuable such referenda are perceived to be. Surely they would not exceed the cost of a presidential election, much less the cost of building an aircraft carrier or of the annual budget for cigarette advertising.

Concluding Comments

In sum, in this chapter we have attempted to identify and clarify some major value issues impending in the further development of the U.S. program for managing radioactive wastes. We offer no recommendations concerning the substantive choices, such as whether to undertake reprocessing of spent fuel assemblies. But we do offer three recommendations concerning the process by which these substantive decisions should be made. First, the federal agency responsible for the program must not take a narrowly technical view of its mission, but should take a broader, value-oriented view that treats public acceptability as a coequal goal with technical containment of radioactivity. Second, instead of trying to persuade the public of an official governmental view—an impossible task—the waste management agency should facilitate debate on many sides of the issues. Third, some kind of decision process must be established that even the loser will find acceptable, and we doubt that anything other than public referenda will be able to accomplish this goal.

We recognize that our proposals are based on factual predictions that could turn out to be wrong and on value preferences that others may not share. Some losers in the radioactive waste controversy may not accept the majority's decisions as easily as we predict. Some winners and nonparticipants may be left even less easy or more fatalistic about nuclear waste than they are now. In our value-laden judgment, however, it is preferable to run the risk of giving too much information and authority to the public than too little. In the long run, we consider it most unlikely that nuclear experts and government officials acting on their own can manage nuclear wastes in a way that will be perceived as legitimate by the public. And we consider it unlikely that public choices on the value issues at the heart of the waste management program will be any worse than the value choices that would be made by elites. Since we consider some waste-management accidents and problems (though not necessarily repository accidents) to be inevitable, the best chance of maintaining public confidence is to prepare citizens with the truth through open debate, to let them make the key value choices, and to leave the way open for further guidance from the public in the future. This is not a prescription for perfection. But perfection is unattainable. Our proposals, we believe, are a basis on which renewed trust can be built among government, industry, environmental groups, and the public.

REFERENCES

Dahl, Robert A. 1970. *After the Revolution? Authority in a Good Society.* New Haven: Yale University Press.

Duderstadt, J.J., and Kikuchi, Chihiro. 1979. *Nuclear Power and Public*

Policy: The Social and Ethical Problems of Fission Technology. Boston: D. Reidel Publishing Co.

Interagency Review Group (IRG). 1978. *Report to the President by the Interagency Review Group on Nuclear Waste Management*. Washington, D.C.: National Technical Information Service. TID-24442. UC-70.

Lipschutz, Ronnie D. 1980. *Radioactive Waste: Politics, Technology, and Risk*. Cambridge, MA: Ballinger.

Lowrance, William W. 1976. *Of Acceptable Risk. Science and the Determination of Safety*. Los Altos CA: William Kaufman, Inc.

Rochlin, Gene. 1976. Irretrievability and Multiplicity: Two Criteria for the Disposal of Nuclear Waste. In *Proceedings of Conference on Public Policy Issues in Nuclear Waste Management*, pp. 130–38. Washington, D.C.: MITRE.

Shrader-Frechette, K. S. 1980. *Nuclear Power and Public Policy: The Social and Ethical Problems of Fission Technology*. Boston: D. Reidel Publishing Co.

INDEX